Environment, Climate Change and International Relations

EDITED BY
GUSTAVO SOSA-NUNEZ
&
ED ATKINS

E-INTERNATIONAL
RELATIONS
PUBLISHING

E-International Relations
www.E-IR.info
Bristol, England
First published 2016

ISBN 978-1-910814-09-3 (paperback)
ISBN 978-1-910814-11-6 (e-book)

Copy Editing: Gill Gairdner
Production: Lauren Ventura
Cover Image: kamchatka via Depositphotos

A catalogue record for this book is available from the British Library

E-IR Edited Collections

Series Editors: Stephen McGlinchey, Marianna Karakoulaki and Agnieszka Pikulicka-Wilczewska

E-IR's Edited Collections are open access scholarly books presented in a format that preferences brevity and accessibility while retaining academic conventions. Each book is available in print and e-book, and is published under a Creative Commons CC BY-NC 4.0 license. As E-International Relations is committed to open access in the fullest sense, free electronic versions of all of our books, including this one, are available on the E-International Relations website.

Find out more at: http://www.e-ir.info/publications

Recent titles

System, Society and the World: Exploring the English School of International Relations (Second Edition)

Restoring Indigenous Self-Determination (new version)

Nations under God: The Geopolitics of Faith in the Twenty-first Century

Popular Culture and World Politics: Theories, Methods, Pedagogies

Ukraine and Russia: People, Politics, Propaganda and Perspectives

Caliphates and Islamic Global Politics (new version)

About the E-International Relations website

E-International Relations (www.E-IR.info) is the world's leading open access website for students and scholars of international politics. E-IR's daily publications feature expert articles, blogs, reviews and interviews – as well as a range of high quality student contributions. The website was established in November 2007 and now reaches over 200,000 unique visitors a month. E-IR is run by a registered non-profit organisation based in Bristol, England and staffed with an all-volunteer team of students and scholars.

About the Editors

Gustavo Sosa-Nunez is Associate Professor with the Research Programme in International Cooperation, Development and Public Policy of the Dr. José María Luis Mora Research Institute (Instituto Mora) in Mexico City. He received his PhD in Politics from the University of East Anglia. His research interests focus on international environmental cooperation, environmental and climate change policies at regional level (European Union and North America) and national level – with air quality and ocean policies both of a particular interest. He is a member of the Mexican Research Network on International Cooperation and Development (REMECID), among other research associations.

Ed Atkins is a PhD Candidate in Energy, Environment & Resilience at the University of Bristol. His doctoral research is focused on the competing perceptions of the environment and, in particular, water and how such understandings interact and compete within discourse – utilising the case study of dam construction in contemporary Brazil. This is with a particular focus on the discourses used to deflect opposition to important schemes of reform and infrastructure construction. His wider research interests include the narratives of climate change, environmental conflict and the Anthropocene.

Abstract

To state that climate change and environment issues are becoming increasingly important in the realm of International Relations is an understatement. Mitigation and adaptation debates, strategies and mechanisms are all developed at the international level, often demonstrating the nuances of international politics and governance. Furthermore, the inherent complexities of climate change make it a particularly difficult phenomenon for international governance. Yet, actions at the international level continue to provide the most effective route to tackle the spectre of climate change.

In the wake of the 2015 Paris conference, this edited collection provides an understanding about the complex relationship between International Relations, the environment and climate change. It details current tendencies of study, explores the most important routes of assessing environmental issues as an issue of international governance, and provides perspectives on the route forward. Each contribution in the collection offers an important understanding of how the Paris agreement cannot be the climax. Rather, as this edited collection shows, it is only the start of global efforts.

Contents

Introduction
Climate Change and International Relations

ED ATKINS

UNIVERSITY OF BRISTOL, UK

&

GUSTAVO SOSA-NUNEZ

INSTITUTO MORA, MEXICO

On the 12th November 2015, a group of fifty students from schools across Bristol, UK, sat down to take part in a semi-structured negotiation exercise with one aim: to piece together a positive global regime to mitigate and adapt to the shifting spectre of climate change.[1] It was a small event but had a significant takeaway. In piecing together international strategies, all participants were forced to wrestle with the complexity of national interests (be they altruistic or self-interested) on the international plain. The result, in short, was chaos. Greenpeace walked out of negotiations. Brazil seized the microphone from the chairs of the session. Not content with Brazil stealing the limelight, Sweden and Russia followed their lead. Significantly, this anarchic nature existed even without the complexities of the climate change regime – there was no right to amendments, no need for consensus and limited pressure from outside of the room. It is a miracle that any form of agreement was found – but it was, and it was overwhelmingly positive.

On the 12th December 2015, delegates of almost 200 countries filed into Le Bourget, Paris for the plenary discussion of the United Nations Framework Convention on Climate Change's (UNFCCC) 21st Conference of Parties (COP-21). After two weeks of frantic – and often nocturnal – negotiations, a draft agreement was reached. Hailed as a historic juncture in the battle against climatic change, this moment possessed a great promise. An aim of keeping temperatures below 1.5 degrees Celsius, the creation of a loss and damages mechanism – although excluding claims of compensation – and the provision of national climate plans all provide an important route forward.

---

[1]   One of the editors of this collection, Ed Atkins, was lucky enough to play a role in the organisation of this event. He would like to extend his thanks to Jack Nicholls, Alice Venn and Chloe Anderson at the University of Bristol – it would be foolish to mention this day without recognising their immense contribution to its success.

Compared to how this conference could have – and previously has – panned out, this was progress. Yet, the critical voices have continued. Vulnerable states and communities appear short-changed, the 1.5-degree target appears unmeetable and the more-progressive mechanisms of mitigation are noticeably absent. Although there was relief in Le Bourget, only time will tell if this is the agreement the world needs.

This book was put together in the weeks and months preceding the Paris conference – and, at the time of writing, the ink of the agreement is not yet dry. As a result, this collection cannot provide any concrete analysis of the route forward that it represents. Instead, it seeks to provide a complementary understanding of how and why the international community must seek to reappraise its understanding of climate change and tactics of mitigation and adaptation. Paris is not the answer – it must only be the start.

This first case above may appear anecdotal – and it most likely is. Thousands of events like it have been run before, across the globe, and they all point to the same conclusion: that not only are climate change and the environment acquiring an increasing importance in the realm of International Relations, but *vice versa*. Climate change mitigation and adaptation strategies and mechanisms are developed at the international scale – resulting in their saturation with the nuts and bolts of international politics and governance. This is problematic: the effects of climate change will not be first experienced at the Westphalian scale of analysis but will instead affect the lives of those at the local level. Thus, it would seem a scalar paradox is created – the phenomena felt at one level and the decision made – or not made – at another.

Yet, it is due to the universal nature of climate change – and responsibilities for it – that the international level provides the most effective route towards mitigation and adaptation (Luterbacher and Sprinz, 2001). Furthermore, the local effects of climate change can transcend the locale and be felt at the international level. The war in Syria – and the refugee crisis associated with it – can be linked to a prolonged period of drought in the country that drove the rural poor into the cities only to find their future limited by authoritarian rule and inaction. Popular protests against hydraulic fracturing ('fracking') and for the divestment from fossil fuel industries have grown into a global coalition of civil society, political parties and individuals. Lastly, the compromised future of many small island developing states (SIDS), due to sea level rise, and the undecided fate of their populations have led to a distinct understanding of the international injustice present within the multilateral negotiations on climate change.

It is this last case that provides us with an important reading of climate change: that it represents an existential crisis. Not for the planet, but for us. Yet, due to the dangerous environmental situations in which many in the world will find themselves, more needs to be done to understand the relations between climate change and international governance. It is this need that this collection takes as its starting point. In doing so, it follows in a rich line of literature. From Paul R. Ehrlich (1971) and Garret Hardin (1968) to Anthony Giddens (2009) and Naomi Klein (2015), the scope of climate change and its governance has captured the minds of many.

Scholars based in security studies have often conceptualised environmental problems as an international security threat – with an increasing emphasis on climate change (see Westing, 1986; Homer-Dixon, 1994; Barnett, 2000). Others have looked to discuss global commons problems (such as ozone depletion and global warming) as issues to be solved in multilateral agreements (see Haas et al., 1993; Yamin and Depledge, 2004). Many perspectives have been provided within this literature, either supporting or rejecting the need for international action. Social Darwinists have theorised a strong connection between nature and the collective humankind (Hofstadter, 1944) that would provide biological justifications of laissez-faire economics in a climate change regime (Leonard, 2009: 38, 40). Within this reading, free markets can perform a crucial role when dealing with natural resource scarcity and degradation, via the provision of incentives. The belief that the market will fix all problems continues. Such activity can be seen in the REDD and REDD+ schemes against deforestation, carbon trading and the privatisation of water supply across the globe. However, it should be noted that such schemes have been widely criticised as a new arena of commodification of resources (Castree, 2003; Swyngdeouw, 2012; Fairhead et al., 2012; Branco and Henriques, 2010). As a result, if we are to reform environmental policies, we should consider both social changes and environmental changes (Rudel, 2013: 5), as well as the manner in which the two interact. It is important to note that climate change forces us all to confront a significant ontological question: what is the essence of our relationship with the natural world?

In short, welcome to the Anthropocene. All organisms transform their habitat to some degree. Woodpeckers make holes in trees, creating sites for nests; rodents burrow; and beavers build dams. However, human society has taken it to a new level. Over half of the planet's large river systems have been fragmented by our dam construction – with over 45,000 large dams disrupting two-thirds of natural freshwater flows across the world. We have drained entire marshes and aquifers. We have altered the carbon cycle, the nitrogen cycle and the acidity of the oceans. We have created urban areas whose dominance and environmental consequences extend well beyond their peripheries. Close to 70 per cent of the world's forests are at a distance of less than half a mile from the forest's edge and the civilisation that exists outside of it. The concept of wilderness is now an historical artefact. The extinction of many species has come as a result of our own actions. Virgin nature has ended; we have harnessed it and constructed our physical environment in such a way that it has become unrecognisable. Gaia is dying and the earth has become a mere footnote in a history of production and consumption.

Climate change is a global commons problem. Its causes – man-made greenhouse gas emissions – and impacts are distributed and felt (albeit not equally) across the international system, transcending traditional boundaries and jurisdictions of the states of the international political system. As a result, causality is particularly difficult to assert in objective terms. Concepts of 'historical responsibility' and 'right to development' are regularly used in the debates surrounding climate change – but their sedimentation is limited. These assertions can be particularly problematic due to the assigning of these concepts to specific states, neglecting the host of non-state actors that operate both within and across state boundaries – all of which share a degree of this responsibility. With these actors important to the story of climate change, it becomes important to understand that states are not monolithic entities but are instead complex groups comprised of small, integrated systems and units that range beyond the realm of international politics. It is this complexity that must be understood when exploring the relationship between climate change and International Relations.

This book aims to provide the reader with an introductory guide to this complexity and the context in which the environment is found and understood in the realm of International Relations. It is important to note that, due to these complexities, it is problematic to base this exploration within a strict framework of International Relations theory. This collection is not for that purpose – it is instead to empirically ground such understanding in the experience of climate change at the international level – be it in the form of conflict, negotiations or the mechanisms created by the global community.

Within this purpose, we have looked to explore what we regard as some of the main topics. We are fully aware that many issues concerning the relationship among the environment, climate change and IR have been left out. Yet, this obeys more to book length concerns, than forgotten issues. Three sections present contributions from authors of diverse academic backgrounds and geographical settings. This is a conscious decision taken at the start of this process. The editors of this book are based in Mexico City (Mexico) and Bristol (UK), respectively, and they have welcomed contributions from five continents. These chapters also represent a range of voices in the academy – from PhD candidates to professors. These two characteristics are important – environmental and climate change issues are global and intergenerational issues and should be treated as such. Lastly, this book embraces the complexity of climate change by weaving interdisciplinary thought throughout – sociology, law and psychology are all represented, allowing a wide exploration of exactly what climate change means in the international sphere.

The book is structured in three distinct sections. The first section, *International Relations Tendencies on Environment and Climate Change*, provides a series of contributions that seek to contextualise this collection, exploring how climate change interacts with the international level. Consisting of five contributions, the section explains the contemporary tendencies that inform international understandings of environmental issues and climate change.

The first chapter, by Mizan R. Khan, introduces the governance of climate change to the theories of International Relations – exploring the role that realism, liberalism and constructivism (among others) play in our understanding of the international regime of climate change adaptation. As Khan asserts, 'climate change is the poster child of global diplomacy today'. Yet, this often ignores the intrinsic complexity of this phenomenon as a policy problem. Khan understands this 'perfect moral storm' (Gardiner, 2006: 33) via a theoretical framework that draws from neoliberalism, regime theory and institutional functionalism, before putting forward a fresh perspective. In doing so, the contribution seeks to explore climate change as creating a new moral norm of global public good and global public bad – opening up analysis of the complexity of climate change adaptation in both theory and practice.

Ursula Oswald Spring provides an account of the complex interrelations and feedbacks between the human system and the environmental. By using an approach that compiles human, gender and environmental security (HUGE security), her contribution explores the viability of multilateral negotiations between governments, business communities and organised societies in relation to long-term sustainability goals. For this, Oswald Spring explains and differentiates concepts like global environmental change and climate change, the 'Anthropocene' and the importance of ecosystem services. Moreover, explanations about the Pressure-State-Response model are provided.

In 'Environment and International Politics: Linking Humanity and Nature', Simon Dalby details the importance that the environment has acquired in International Relations scholarship, the debates around it, and the nascent links it has with security, peace, and war. Dalby also explains the different understandings the term 'environment' produces, both politically and in materiality, to actors in both the North and the South. The role of science in major international events both during and after the Cold War demonstrates the important role that the environment has when prompting international action of any kind. In addition, Dalby provides further insights into the importance of international agreements, environmental security, political economy and climate change, as well as where the future lies in the 'Anthropocene'.

Nina Hall looks at current trends to argue that climate change has become institutionalised in global affairs as a top priority issue, identifying four dimensions that confirm this: scientific consensus, political action, the location financial resources and the institutionalisation of climate change multilateral organisations. Hall examines G7 and G8 communiqués as well as international organisations' engagement with the United Nations Framework Convention on Climate Change (UNFCCC). This approach allows the concluding observation that, although climate change was previously minimised by international actors, this trend is reversing.

In the final contribution to this section, Kirsti M. Jylhä explores the psychology of humans' reluctance to acknowledge climate change as a man-made problem. Jylhä suggests a move from questioning whether climate change is caused by humans to asking what hinders people from acknowledging it as an important route for research. In doing so, she affirms that denial develops for many different reasons, within a range of psychological mechanisms. In addition, Jylhä relates the concept of Social Dominance Orientation (SDO) to processes of climate change denial, stating that the perception that humans have of themselves as a superior group tends to compound the perceived right to dominate the rest of the nature.

The second section of the book, titled *Assessments: Which Way to Follow?*, presents the reader with four different contributions that explore the manner in which we – as an international community – understand environmental problems, climate change implications and the policy mechanisms that are in existence.

In 'Transversal Environmental Policies', Gustavo Sosa-Nunez presents an insight into the role environmental policies may play within a wider policy framework. This transversal nature is noticeable, but their omission or partial involvement is also obvious. In this context, Sosa-Nunez comments on policy approaches to environmental management; listing administrative rationalism, democratic pragmatism and economic rationalism as options through which inclusion of the environment in broader policies can take place. Furthermore, Sosa-Nunez addresses the role that environmental policies have in broader policy frameworks. This is illustrated through different areas such as industry, security, science, climate change and urban planning. Sosa-Nunez then goes on to identify the adequate conceptualisation of environmental policies – questioning whether transboundary cooperation or international governance could better explain the transversal approach that environmental policies have.

In his contribution, Ed Atkins explores the widely cited spectre of environmental conflict. Within this reading of degradation and change, many have asserted that a chain of causality will develop, with environmental pressures leading to increased competition that results in conflicts over scarce resources. This contribution looks to debunk this assertion, arguing for the drawing of an important distinction between strictly environmental factors and resources of an economic nature. It is the latter that provides an important understanding of the Anthropocene – with society's interaction with these resources (such as oil and gas) bestowing value upon them, driving potential competition. Instead, this contribution argues for a focus on strictly environmental routes to conflict. This opens up analysis to the role of the environment in a wider causal web of conflict – as demonstrated in the case of conflicts over environmental refugees.

In her contribution, Emilie Dupuits affirms that the increased participation of non-state actors in international governance is occurring due to the high fragmentation in which global environmental governance is found. This, Dupuits claims, is an opportunity for civil society and non-governmental organisations. However, she also recognises that this possibility leads to competition for visibility and power, which can hamper the strength of participation. By revising literature on multi-scalar governance, Dupuits asserts the importance of state and non-state actors in the transition from a hierarchical international system towards a horizontal network.

In 'Global Climate Change Finance', Simone Lucatello engages in the debate about who is going to pay for mitigation and adaptation costs within national and international responses to climate change. In doing so, he explores the effectiveness of environmental aid and economic initiatives. Lucatello suggests that multilateral aid is preferable to bilateral aid for a number of reasons. First, it provides greater financial control to recipient, generally developing, countries. Second, a multilateral scheme is more desirable because it is less open to political issues and can be better delivered, therefore providing better outcomes. However, issues remain over the origin of economic resources. Who should pay, how should the money be delivered and what should its destinations be are questions that ought to accompany concerns surrounding climate change governance.

The third section of the book, *Two Steps Forward, One Step Back: Perspectives as We Continue with Our Lives*, provides an insight into the actions and processes we should expect of international environmental governance in the future. This section seeks to take into account the problems we, as an international community, face if we are to find and maintain resilience in the face of environmental problems and climatic change. It is important to note, as is outlined in this section, that much of this will depend on the commitment that different sectors of national and international society can provide to ensure this resilience for forthcoming generations.

In his chapter, Lau Blaxekjær explores the emergent role of the study of environmental diplomacy as an additional lens of understanding within International Relations literature. Using the examples of the role of the Cartagena Dialogue in UNFCCC negotiations and the influence of green growth networks, Blaxekjær posits that contemporary scholarship must seek to understand the 'orchestrating role that diplomacy plays in these new, overlapping environmental governance fields' present within the climate change regime. It is important to note, as this contribution does, that these coalitions often take the form of partnerships, utilising tactics of issue linkage. With the international governance of climate change standing at an important crossroads within the post-Paris regime, it is important to explore the increasing role of these partnerships in the development of the international relations of climate change.

As Duncan Depledge explores in his analysis of the geopolitics of the region, the observed and predicted climatic changes will be particularly experienced in the Arctic Circle – a region which overlaps the territorial boundaries of a number of states, including the USA and Canada, the Scandinavian states and Russia. Depledge charts how this has resulted in decisions over the Arctic occurring at all levels of governance – from the community to the global. A wide-ranging discussion follows regarding the best route forward and how it should be taken. A particular issue that has become significant within these processes has to do with wider understandings of political economy: will the Arctic provide a new resource frontier or a global commons?

In 'Renewable Energy: Global Challenges', Lada V. Kochtcheeva explores the inherent complexity of piecing together the implementation of renewable energy strategies. Although the use of renewable energy is increasing across the globe, the success of these measures – and their wider adoption – are often constrained by a series of regulatory, technological, social and economic barriers. As Kochtcheeva argues, this is often the result of the need to balance competing policy goals – such as sustainability and economic development. Large-scale subsidies for fossil fuels and nuclear power persist – often resulting in the undercutting of renewable energy technology. These market failures can often be coupled with unfavourable institutional environments that further limit successful adoption. This contribution argues that the solution could be found in a more systematic approach of research, one that aims to further our understanding of the unfavourable conditions that hinder the adoption of renewable energy.

The final two chapters can be understood as a twinned approach to a fast-growing actor in the international governance of climate change: the fossil fuel divestment campaign. While the individual demands of the movement may vary on the ground, the overarching aim is simple: that companies and institutions divest (or withdraw investment) from companies that profit from the fossil fuel industry (such as oil and gas companies). First, prominent activists in the USA Leehi Yona and Alex Lenferna look to the student-led divestment movement as a means to understand the future of popular understandings of climate change and its interaction with international society. The movement has its roots in a 2010 campaign at Swathmore College, Pennsylvania, and has transformed extensively since then. Yona and Lenferna argue that this is the result of a process of coalition-building, support/pressure from alumni groups and the transformative generational belief that you cannot solve the problem by supporting the actors that created it.

In the final chapter, Matthew Rimmer presents an important, primary source-laden analysis of how this divestment movement has also sought to influence the management of sovereign wealth funds. Using the example of Norway, Rimmer explores the way that the popular divestment movement has globalised its efforts – striking at the heart of the international system. The bold decision by the Norwegian government to divest from the coal industry can be used as an example for many nations to follow. Rimmer argues that, although the introduction of divestment as a policy initiative at the international level remains uncertain, its future role will likely present important options in the international climate law regime.

To conclude, the editors present what they consider are the key findings of this edited collection. They offer a critical assessment on the context of the environment and climate change within IR studies, before concluding with suggestions for the development of future understanding of the mutually constitutive relationship between climate change and International Relations.

References

Barnett, J. (2000). Destabilizing the Environment–Conflict Thesis. R*eview of International Studies*, 26(2), 271-288.

Branco, M. and Henriques, P. D. (2010). The Political Economy of the Human Right to Water. *Review of Radical Political Economics*, 42(2), 142-155.

Castree, N. (2003). Commodifying What Nature? *Progress in Human Geography*, 27(3), 273-297.

Ehrlich, P. (1971). *The Population Bomb*. London: Macmillan.

Fairhead, J., Leach, M. and Scoones, I. (2012). Green Grabbing: A New Appropriation of Nature. *The Journal of Peasant Studies*, 39(2), 237-261.

Gardiner, S. M. (2006). A Perfect Moral Storm: Climate Change, Intergenerational Ethics and the Problem of Moral Corruption. *Environmental Values*, 15, 397-413.

Giddens, A. (2009). *The Politics of Climate Change*. Cambridge: Polity Press.

Haas, P. M., Keohane, R. O. and Levy, M. A. (eds) (1993). *Institutions for the Earth: Sources of Effective International Environmental Protection*. Cambridge, MA: MIT Press.

Hardin, G. (1968). The Tragedy of the Commons. *Science*, 162(3859), 1243-1248

Hofstadter, R. (1944). *Social Darwinism in American Thought*. Philadelphia: University of Pennsylvania Press.

Homer-Dixon, T. F. (1994). Environmental Scarcities and Violent Conflicts: Evidence from Cases. *International Security*, 19(1), 5-40.

Klein, N. (2015). *This Changes Everything: Capitalism vs. The Climate*. London: Penguin Books.

Leonard, T. C. (2009). Origins of the Myth of Social Darwinism: The Ambiguous Legacy of Richard Hofstadter's Social Darwinism in American Thought. *Journal of Economic Behavior & Organization*, 71, 37-51.

Luterbacher, U. and Sprinz, D. F. (2001). *International Relations and Global Climate Change*. Cambridge, MA: MIT Press.

Rudel, T. K. (2013). *Defensive Environmentalists and the Dynamics of Global Reform*. Cambridge: Cambridge University Press.

Swyngedouw, E. (2012). UN Water Report 2012: Depoliticizing Water. *Development and Change*, 44(3), 823–835.

Westing, A. H. (1986). *Global Resources and International Conflict: Environmental Factors in Strategic Policy and Action*. Oxford: Oxford University Press.

Yamin, F. and Depledge, J. (2004). *The International Climate Change Regime: A Guide to Rules, Institutions and Procedures*. Cambridge: Cambridge University Press.

SECTION I:

INTERNATIONAL RELATIONS TENDENCIES ON ENVIRONMENT AND CLIMATE CHANGE

1

Climate Change, Adaptation and International Relations Theory

MIZAN R. KHAN
NORTH SOUTH UNIVERSITY, BANGLADESH

Climate change is the poster child of global diplomacy today. In fact, it can easily be regarded as the most complex global policy problem. This complexity in understanding the political economy of climate change is reflected in its temporal, spatial and conceptual dimensions. It's a stock rather than a flow pollution problem. Historical emissions from industrial countries are mixing with today's rapidly growing emissions from the developing countries. The impacts will manifest themselves fully in the decades to come, and future generations are likely to suffer the most; yet scientists already attribute the trend of increased magnitude, frequency and severity of climate disasters of recent years to climate change (IPCC 2012). The main creators of the problem are the rich industrial countries, which are likely to suffer less; while the poor, with the least contribution to the problem, will suffer the most.

The conceptual dimension of adaptation is much more complex. Climate change is global in both its cause and effect dimensions. As climate change is really a collective action problem, there is a built-in compulsion for addressing the root causes through international cooperation. The mitigation regime is not yet succeeding because of disagreements over cost or the sharing of responsibility among the parties to the United Nations Framework Convention on Climate Change (UNFCCC), but nobody questions the properties of a stable climate as a life-support 'global public good' (GPG). This has been reflected in the Durban Platform agreed at COP17 in

December 2011, which stipulates that all UNFCCC parties have to accept mitigation responsibility.

Gardiner (2011: 398) aptly calls the climate change problem a 'perfect moral storm', at the base of which lies his thesis of 'theoretical ineptitude' (p. 407). In this chapter we argue that the alleged lacuna lies more in conceptualising adaptation. To do this, we turn to the main theories of International Relations, such as realism, regime theory, neoliberalism, and constructivism, to see how climate change and adaptation are viewed by these strands. In international relations, a state can take any of the three approaches: cooperation, unilateralism or inactivity. Within the realm of climate diplomacy, we witness states playing all these roles.

Realism is perhaps the most influential strand in International Relations, particularly during the Cold War, to have guided nations in their foreign policy pursuits. The central premise of this theory is that in an anarchic space with no order, nations are guided as unitary rational actors by maximising interests based on power politics. In this pursuit countries employ the mechanisms of power at their disposal to turn the deals in their favour. To realists or rational choice theorists, ethics, moral values and justice have no place in international politics and are instead viewed as 'oxymoronic expression[s]' (Franceschet, 2002; Okereke, 2010). Vanderheiden (2008) argues that realist theory, through a prism of only looking at national interests, may show concern with increasing global poverty due to the perception that this may increase security threats rather than any injustice endemic to global poverty itself. Likewise, a realist understanding might support a climate treaty with mandatory limits to greenhouse gas (GHG) emissions if national interests are better served with these than without. This might also be the case with assistance in adaptation to developing countries.

The Copenhagen Accord, worked out by the leaders of Brazil, China, India, South Africa and the United States (US), is viewed as a return to realism, though some scholars disagree (Bernstein et al., 2010). Though the main concern of the Copenhagen Accord architects is mitigation, it contains rich references to adaptation. Two points may be mentioned: first, the urge for international cooperation for adaptation and, second, the need for a balanced allocation of the pledged amount of US$30 billion between adaptation and mitigation. Vanderheiden (2008) further posits that the effects of climate change on other people with no spill over effect on a realist do not bother him. From this perspective, adaptation in developing countries is not a concern for rich states since it does not provide them with any direct benefit (Barrett, 2008). In contrast to this perspective, normative international political theory brings the issue of international justice into focus. Brown argues that normativism emphasises that states will act not just for self-interest but also in accordance with justice-related principles, whereby 'states receive what is their *due* or have the *right* to expect certain kinds of treatment' (Brown, 2002: 276).

Liberalism and its later version, neo-liberalism, argue that nations benefit from cooperation in an atmosphere of peace and harmony. Former US president Woodrow Wilson was a premier advocate of liberalism. Along these lines, some argue that without funding for adaptation, many vulnerable developing countries might not remain viable partners in trade and investment. Further, climate-induced migration may engender conflicts within and across regions. With this understanding, adaptation funding is viewed as inducing developing countries to go for mitigation (Buob, 2009). Self-interest dictates that industrial countries should provide funding for adaptation.

Significantly, the core elements of the UNFCCC and the Kyoto Protocol reflect the economic orthodoxy of neoliberalism, i.e. the level of acceptable GHG concentration is determined through cost-benefit analysis. To achieve this level with least cost, market mechanisms are required (Article 3.3 of the Convention, and Articles 6, 12 and 17 of the Kyoto Protocol). Adaptation concerns present a poor case to be taken care of by market-based instruments (Barrett, 2008). Driesen (2009) argues that barriers to promoting adaptation concern the free market orthodoxy under the neoliberal agenda worldwide, with markets, not governments, ruling the game – as in the way that atmospheric sink capacity has been turned into property rights through carbon trading (Newell and Paterson, 2010). More on this follows in the last section.

Regime theory argues that nation-states are the central actors in global negotiations, with civil society playing only a minor or supportive role in shaping outcomes. Regimes are defined as sets of principles, norms, rules and decision-making procedures around which actor expectations converge in a given issue area (Krasner, 1982). Young, Keohane and Nye are leading advocates of regime theory (Keohane, 1989; Nye, 1991). As climate change is a global phenomenon, regime theorists focus on mitigation rather than adaptation. The climate regime reflects this strand, though talks of increasing cooperation about adaptation are present. This is due to the mutuality of interests in mitigation. Actually, regime theory reflects the values of liberal institutionalism, which considers international institutions to be a force in global politics. For environmental problems straddling the global commons, it is difficult to draw a dichotomy, as statist model does, in policy debates between domestic and international sphere, and it is in these common issues that international organisations play an active role. For this reason, Rosenau (1997) challenged the statist model in his work on global governance. This is true particularly in climate change diplomacy, as the UNFCCC Secretariat, the United Nations Environment Programme (UNEP), the United Nations Development Programme (UNDP), the World Bank and some other bodies play very important roles in articulating and setting the agenda for discussion.

In their book, Bulkeley and Newell (2010) present a critique of this power-based regime theory. According to them, regimes are formed and dominated by a hegemon. Unlike power-based accounts, functionalists of interest-based approaches to regimes are concerned with how different institutional designs shape and affect the behaviour of nations. Along these lines, a political economy critique states that these institutions, with the agenda of promoting neo-liberal market philosophy, help capital formation and perpetuate the existing order. Tanner and Allouche (2011) argue that within a liberal-market system, climate change is seen as a challenge that threatens to derail progress in poverty reduction and the dominant mode of capitalist development. Newell and Paterson (1998) argue that, as a result of corporate power, international capital's response to climate change is weak.

Compared to regime theory's 'high politics' approach to international relations, political ecology brings in the 'low politics' issues of global politics, such as inequality, poverty, structural weaknesses and the ethical and justice dimensions, including compensation for damages around which the climate change debate is centred (Jamieson, 2001; Adger et al., 2006; Roberts and Parks, 2007; Okereke, 2008; Abdullah et al., 2009). Saurin (2001) argues that non-recognition of political ecology considerations in climate change is hardly surprising and this is reflected in ignoring scholars writing about social, political and economic conditions because they are largely unconcerned with the state system. Thus, political ecology is viewed as presenting an alternative to conventional analyses of the climate regime by its way of explaining economic rationality through social and environmental lenses (Glover, 2006). It is concerned more with the implications of Convention outcomes for ecological justice among present and future generations and for non-human life, and also with applying the 'Commons' concept to the global atmosphere (Agarwal and Narain, 1991; Shue 1992, 1999; Byrne 1997; Volger, 1995; Brown 2002). Singer (2004) argues that national boundaries, in their traditional conceptualisations, are rendered obsolete by global environmental problems such as climate change.

Constructivism finds its origins in a challenge to positivism that focuses on the epistemological perspective – i.e. that the nature of scientific knowledge is 'constructed' by the scientists (Kincheloe, 2005). While the physical sciences employ descriptive paradigms with quantitative tools, social science research is often conducted within an interpretive paradigm, which focuses on the meaning people ascribe to various aspects of their lives based on cultural values (Rayner and Malone, 1998). As Kuhn (1970) stated, what a man observes depends upon what his previous visual-conceptual experience has taught him to see. So, this method argues that reality is subjective and that 'truth' is therefore a construction reflecting our own experiences – historical, cultural and experiential. And this interpretation is not static but dynamic, evolving over time as a result of interactions with other peoples and entities.

In International Relations, constructivists emphasise a shift away from rationalist and interest-based accounts to factor in the role of knowledge, norms and values in shaping positions adopted by nation-states; and see cooperation among nations as guided not just by material and power factors but also by discursive power and ideational elements (Haas et al., 1993; Okereke, 2010). As evidence of discursive power in inter-state relations, Cox (1981) argues that the US rise to and reproduction of global dominance in the 20th century was due to its blending of material and discursive power. The constructivist accounts point to their position by indicating at the intergovernmental panel on climate change (IPCC) epistemic communities, which continue to shaping the climate agenda, with their periodic scientific assessments. Constructivist scholars focus more on the discursive and intersubjective procedures by which international governance develops (Ruggie, 1998).

Somewhat similar, but another strand by name, **cosmopolitanism** calls for a global order based on justice, human rights and international law (Held, 2009); one in which non-state actors play an increasingly important role. This school argues that, due to globalisation, human beings are bound together and that the vital basic needs of global communities should be prioritised over trivial ones (Shue, 1992; 1999). Under the formulations of constructivism, it can be argued that since adaptation has not been defined or conceptualised in a coherent manner in the climate regime, there is an active process of knowledge-building in adaptation science and policy design, as well as implementation. Along this line of new norm setting and strengthening, adaptation is argued to be a global public good (GPG).

New norm of adaptation as a global public good

The nature of the global public good entails two basic properties: non-excludability and non-rivalness. The former denotes that nobody can be excluded from using a resource, while the latter says that use by one person or one country will not reduce the quantity or quality of a resource for another. It is worth noting that nothing is inherently excludable – policies or social institutions are required to make any good or service excludable. On the other hand, some goods/services are inherently non-excludable as a physical characteristic (Karlsson-Vinkhuyzen et al., 2012). One example is climate regulation. It is also important to note that rivalness is a physical characteristic of a good or service and is not affected by human institutions.

However, climate stability or atmospheric sink capacity may be better conceptualised as a common pool resource (CPR), which is rivalrous; many environmental resources including atmospheric sink capacity can more accurately be described as CPR (Barkin and Shambaugh, 1999). This rivalness is a source of power for those in negotiations and unwilling to replenish the CPR (DeSombre, 2000). From the moment anthropogenic climate change and its negative impacts were first detected by scientists, the atmospheric sink could no longer be regarded as a pure public good because it remains non-excludable. Hence, it can be regarded as a 'congestible public good'. Or better, it can be termed as a global commons, with a finite capacity to absorb atmospheric pollution. The IPCC and other studies, including the US National Assessment, have persistently been trying to convey this message to the world community (IPCC, 2012). So climate change is rightly regarded as the classic case of Hardin's 'Tragedy of the Commons' (1968), while Stern calls it the greatest market failure of our time (2007). The latter happens when the market does not factor in the externality cost and imposes it on society. From the perspective of the prisoner's dilemma, the collective good of potential cooperation, compared to the collective bad, usually makes cooperation possible (DeSombre, 2000); however, the mainstream conceptualisation of adaptation has continued, largely narrow interest- and discipline-based.

Even within the traditional paradigm of thinking, funding for adaptation can bring in direct or indirect global benefits, such as better monitoring and prediction of climate change, improved modelling of climate impacts, research and development (R&D) to improve drought and flood-resistant crops, etc. Also adaptation measures may prevent climate-induced displacement, regarded as an indirect global benefit (Pickering and Rubbelke, 2014).

Accordingly, a number of scholars have started theorising the normative aspects of allocating funds for adaptation from multilateral sources (Paavola and Adger, 2005). Others are looking at adaptation funding as a way to induce the development of mitigation strategies (Buob, 2009). A few studies have discussed the use of vulnerability indices for countries as a basis for distributing climate funds (Klein, 2010). Other studies have started exploring various metrics for comparing the effectiveness of climate change adaptation projects (Stadelmann et al., 2011). Some others have started talking about the emergence of a global governance of adaptation (Otterstrom and Stripple, 2012). However, none of these initiatives attempt to conceptualise climate impacts in terms of failed mitigation as a global public bad (GPB), so taking care of the consequences through adaptation as a GPG. Vanderheiden's idea of adaptation appears expansive, tending to plug the conceptual gap a little: 'Adaptation intervenes in the causal chain between climate change and human harm, allowing the former but preventing the latter, but when this is not possible, a third category of *compensation* costs must be assigned in order to remedy failed mitigation and adaptation efforts [...] so adaptation shall be understood to include prevention of harm as well as *ex post* compensation to it' (2011: 65).

Together, the works of Kaul et al. (1999; 2003) on GPGs under the UNDP banner are important in terms of their new and expanded interpretations. With the onset of globalisation, they bring in both goods and bads (i.e. enhanced economic growth and trade, and widening disparity and growing negative externalities). They argue that a new understanding of a global public good that is different from the conventional national public goods under neoclassical interpretations is needed. The UNDP proposed a broader definition, integrating three elements, called the triangle of publicness: a) publicness in consumption, b) publicness in distribution of benefits, and c) publicness of decision-making. Kaul (2013: 133) defines GPGs as 'goods whose benefits or costs are of nearly universal reach or which potentially affect anyone anywhere. Together with regional public goods they constitute the category of transnational public goods'. Kaul et al. (1999) classified various types of GPG into three groups: a) global natural commons, such as high seas and the atmosphere, b) global human-made commons, such as global networks, knowledge and international regimes, and c) global policy outcomes and conditions, such as peace, security and financial stability.

Sweden and France are regarded as pioneers in embracing the GPG approach (Kaul et al., 1999), and these two countries established an international task force on GPGs in early 2003. This group (International Task Force on Global Public Goods, 2006) defined GPGs as issues that are considered important to the global community, which cannot be provided by individual countries acting alone, and which must be addressed collectively by both developed and developing countries. Along these lines, this task force, together with others, identified tackling climate change as a GPG and included strategies, such as strengthening adaptive and supporting capacity-building in developing countries. The World Bank commissioned a study of its own, looking at its role in the provision of GPGs (Evans and Davies, 2015). This broadened concept of GPG was based on the fusion of several theoretical strands: a) the theory of public goods, as understood in economics, 2) the theory of market failure, in terms of positive and negative externalities, c) the theory of basic needs, to justify the notion of free access to resources, and d) elements of political economy, to define collective actions and collective goods (Kaul et al., 2003: 185). However, such an expanded interpretation of GPGs has its critics at both academic and policy level. For example, Long and Woolley (2009) argue that the UN interpretation of GPGs is rhetorically effective but poorly defined, lacking conceptual clarity and with too many abstractions. Furthermore, they argue that the 'concept gives a simple rationale for the activities of those associated with UN agencies [...] to fit the exigencies of international public policy rather than explanatory theory' (Long and Woolly, 2009: 118). At the policy level, there are both GPG supporters, such as the European Union (EU) countries, and opponents, like Japan and the US. The central issues that differentiate them are the interrogations of additionality of financial resources, over and above foreign aid. Developing countries feared the diversion of official development assistance (ODA) to GPG provision (without additionality) (Carbone, 2007).

However, this thinking is no longer justifiable in an era of growing commons problems accompanied by rapid and uneven globalisation. The traditional understanding of GPGs as national and territorial is called into question by this new crop of extraterritorial problems. Cross-border externality problems now represent a group of GPBs, warranting their collective internalisation into national and global policy processes. Even the widening disparity and concentration of poverty in the middle-income countries is now viewed by some as a GPB, meriting a collective solution. In the case of climate impacts and adaptation, the critiques can be refuted in a number of ways: first, a deeper analysis will reveal that adaptation benefits extend from the national to the global level, both directly and indirectly (Table 1, below), and ambitious mitigation strategies bring in adaptation benefits in the form of avoided loss and damage. But this is not taking place. Vanderheiden argues that adaptation must include both the prevention of harm and *ex post* compensation for unavoidable loss and damage. Moreover, norms such as human rights, the right to development and the no-harm rule are globally recognised and regarded as a new class of GPGs. Obviously, both mitigation and adaptation appear as important GPGs to ensure the realisation of related norms. Volger (1995) talks of the shared vulnerability or global fate interdependence that climate change has engendered. Instead of exercising the centuries-old Westphalian, realism-based concept of sovereignty, a new type – what Kaul (2013) calls smart or pooled sovereignty – is warranted for addressing this new type of transnational problem. Finally, let us have a look at the multidimensional and multilevel benefits of adaptation. The table below shows the types of benefits, with examples, along three dimensions: whom they accrue to (private/public), their geographic scale (local to global), and whether they are of a direct or indirect nature.

Table 1: Key types of adaptation benefits

Local private benefits	Local public benefits	Direct global public benefits	Indirect global public benefits
Value of saved crops for individual farmers; improved water storage for households.	Flood-proofed infrastructure, afforestation preventing mudslides, coastal afforestation as wind and flood breaks, water storage.	Control of climate-sensitive infectious diseases, protection of climate-sensitive biodiversity, agricultural research on flood and saline-resistant crops, improved modelling of climate impacts.	Continuance of statehoods by many small island states, avoided international migration, lower price volatility on climate-sensitive agricultural products, enhanced purchasing power among the vulnerable communities and countries.

Source: Adapted from Persson (2011) and expanded by the author.

The list thus amply manifests that adaptation, jointly with its diverse and multi-level benefits, does contribute to both direct and indirect global benefits. Central to this articulation are social constructivism and normative international political theory, which argue that questions about norms, morality and justice are not external but very much intrinsic to interactions between states in the 21st century (Shue, 1992; Franceschet, 2002; Okereke, 2010).

Conclusion

This chapter has reviewed the main strands of International Relations theory, such as realism, liberalism, regime theory and constructivism, in order to see how they approach global cooperation in adaptation. The review shows that all strands have elements of cooperation for adaptation, but with varied ways and perspectives. The current climate regime generally reflects a mix of neoliberalism, regime theory and institutional functionalism. However, in accordance with Einstein's argument that the solution of a problem requires rising above the level of consciousness that created it, this chapter follows evolving constructivist thinking, preparing the ground for the advent of a new norm – an expanded interpretation of GPG/GPB in an era of increasing global commons problems. Such an exercise has the potential to command a more robust political response to globalising the responsibility for addressing adaptation. Though this new norm of considering adaptation as a GPG is in its embryonic stage, it can be expected that there will be further conceptualisations by the theorists of governing global commons such as atmospheric sink capacity.

References

Abdullah, A., Muyungi, R., Jallow, B., Reazuddin, M. and Konate, M. (2009). *National Adaptation Funding: Ways forward for the Poorest Countries*, IIED Briefing Paper. London: International Institute for Environment and Development.

Adger, W. N., Paavola, J., Huq, S. and Mace, M. J. (eds) (2006). *Fairness in Adaptation to Climate Change*. Cambridge, MA: MIT Press.

Agarwal, A. and Narain, S. (1991). *Global Warming in an Unequal World: A Case of Environmental Colonialism*. New Delhi: Centre for Science and Environment.

Barkin, J. S. and Shambaugh, G. E. (eds) (1999). *Anarchy and the Environment: The International Relations of Common Pool Resources*. New York: State University of New York Press.

Barrett, S. (2008). Climate Treaties and the Imperative of Enforcement. *Oxford Review of Economic Policy*, 24(2), 239-258.

Byrne, J. (1997). *Equity and Sustainability in the Greenhouse: Reclaiming our Atmospheric Commons*. Pune, India: Parisar.

Brown, D. A. (2002). *American Heat: Ethical Problems with the United States' Response to Global Warming*. Lanham, MD: Rowman & Littlefield.

Bulkeley, H. and Newell, P. (2010). *Governing Climate Change*. Abingdon: Routledge Global Institutions.

Buob, S. (2009). *On the Incentive Compatibility of Funding Adaptation*. Working Paper 2009/03. Bern, Switzerland: National Centre of Competence in Research on Climate (NCCR Climate).

Carbone, M. (2007). Supporting or Resisting Global Public Goods? The Policy Dimension of a Contested Concept. Global Governance: A *Review of Multilateralism and International Organizations*, 13(2), 179-198.

Cox, R. W. (1981). Social Forces, States, and World Orders: Beyond International Relations Theory. *Millennium Journal of International Studies*, 10(2), 126-155.

DeSombre, E. R. (2000). Developing Country Influence in Global Environmental Negotiations. *Environmental Politics*, 9(3), 23-42.

Driesen, D. M. (ed.) (2009). *Economic Thought and U.S. Climate Change Policy*. Cambridge, MA: MIT Press.

Evans, J. W. and Davies, R. (eds) (2015). *Too Global to Fail. The World Bank at the Intersection of National and Global Public Policy in 2025*. Washington, DC: IBRD/The World Bank. Retrieved from https://openknowledge.worldbank.org/bitstream/handle/10986/20603/928730PUB0Box3021030709781464803079.pdf?sequence=1

Gardiner, S. M. (2011). *A Perfect Moral Storm: The Ethical Tragedy of Climate Change*, Oxford: Oxford University Press.

Franceschet, A. (2002). Moral Principles and Political Institutions: Perspectives on Ethics and International Affairs. *Millennium Journal of International Studies*, 31(2), 347-357.

Glover, L. (2006), *Post-modern Climate Change*, London & New York: Routledge.

Held, D. (2009). Restructuring Global Governance: Cosmopolitanism, Democracy and the Global Order. *Millennium Journal of International Studies*, 37(3), 535-547.

Hardin, G. (1968). The Tragedy of the Commons. *Science*, 162(3859), 1243-1248.

Haas, P. M., Keohane, R. O. and Levy, M. A. (eds) (1993), *Institutions for the Earth: Sources of Effective International Environmental Protection*. Cambridge, MA: MIT Press.

International Task Force on Global Public Goods (2006). *Summary: Meeting Global Challenges: International Cooperation in the National Interest.* Washington, DC.: Communications Development Incorporated.

IPCC (2012). *Managing the Risks of Extreme Events and Disasters to Advance Climate Change Adaptation.* Special Report. Cambridge: Cambridge University Press. Retrieved from https://www.ipcc.ch/pdf/special-reports/srex/SREX_Full_Report.pdf

Jamieson, D. (2001). Climate Change and Global Environmental Justice. In: P. Edwards and C. Miller (eds). *Changing the Atmosphere: Expert Knowledge and Global Environmental Governance* (pp. 287-307). Cambridge, MA: MIT Press.

Kaul, I. (2013). Accelerating Poverty Reduction through Global Public Goods. In: OECD. *Development Co-operation Report 2013: Ending Poverty* (pp. 131-139). Paris: OECD Publishing. Retrieved from http://www.oecd-ilibrary.org/docserver/download/4313111ec017.pdf?expires=1449549049&id=id&accname=guest&checksum=CAD567A41B4DEBC6358659F7F7D65CB8

Kaul, I., Grunberg, I. and Stern, M. A. (eds) (1999). *Global Public Goods: International Cooperation in the 21st Century.* New York: Oxford University Press.

Kaul, I., Conceicao, P., Le Goulven, K. and Mendoza, R. U. (eds) (2003). *Providing Global Public Goods: Managing Globalization.* New York: Oxford University Press.

Klein, R. J. T. (2010). Mainstreaming Climate Adaptation into Development: A Policy Dilemma. In: A. Ansohn and B. Pleskovic (eds). *Climate Governance and Development*, Berlin Workshop Series 2010 (pp. 35-52). Berlin: World Bank.

Keohane, R. O. (ed.) (1989). *International Institutions and State Power: Essays in International Relations* Theory. Boulder, CO: Westview Press.

Kincheloe, J. L. (2005). *Critical Constructivism.* New York: Peter Lang Primer.

Kuhn, T. S. (1970). *The Structure of Scientific Revolutions.* Chicago: University of Chicago Press.

Karlsson-Vinkhuyzen, S. I., Jollands, N. and Staudt, L. (2012). Global Governance for Sustainable Energy: The Contribution of a Global Public Goods Approach. *Ecological Economics*, 83, 11-18.

Krasner, S. D. (1982). Structural Causes and Regime Consequences: Regimes as Intervening Variables. *International Organization*, 36(2), 185-205.

Long, D. and Woolley, F. (2009). Global Public Goods: Critique of a UN Discourse. *Global Governance: A Review of Multilateralism and International Organizations*, 15(1), 107-22.

Newell, P. and Paterson, M. (1998). A Climate for Business: Global Warming, the State and Capital. *Review of International Political Economy*, 4(5), 679-704.

Newell, P. and Paterson, M. (2010). The Politics of the Carbon Economy. In: M. T. Boykoff (ed.). *The Politics of Climate Change* (pp. 77-95). London: Routledge.

Nye, Jr., J. S. (1991). *Bound to Lead: The Changing Nature of American Power*. New York: Basic Books.

Okereke, C. (2008). *Global Justice and Neoliberal Environmental Governance: Ethics, Sustainable Development and International Cooperation*. Abingdon: Routledge.

Okereke, C. (2010). Climate Justice and the International Regime. *Wiley Interdisciplinary Reviews*: Climate Change, 1(3), 462-474.

Otterstrom, G. D. and Stripple, J. (2012). *Legitimacy in Global Adaptation Governance*. Lund: Earth System Governance/Lund University. Retrieved from http://www.earthsystemgovernance.org/lund2012/LC2012-paper197.pdf

Rayner, S. and Malone, E. L. (1998). The Challenge of Climate Change to Social Sciences. In: S. Rayner and E. L. Malone (eds). *Human Choice and Climate Change: An International assessment*. Vol. 4: What Have We Learned? Columbia, OH: Battelle Press.

Persson, A. (2011). *Institutionalising Climate Adaptation Finance Under the UNFCCC and Beyond: Could an Adaptation 'Market' Emerge?* Stockholm Environment Institute Working paper 2011-03. Retrieved from http://www.sei-international.org/mediamanager/documents/Publications/Climate/Adaptation/sei-wp-2011-03-adaptation-commodification.pdf

Paavola, J. and Adger, W. N. (2005). Fair Adaptation to Climate Change. *Ecological Economics*, 56, 594-609.

Pickering, J. and Rübbelke, D. (2014). International Cooperation on Adaptation to Climate Change. In: A. Markandya, I. Galarraga and E. Sainz de Murieta (eds). *Routledge Handbook of the Economics of Climate Change Adaptation* (pp. 56-75). Abingdon: Routledge/Earthscan.

Roberts, T. J. and Parks, B. C. (2007). *A Climate of Injustice: Global Inequality, North-South Politics, and Climate Policy.* Cambridge, MA: MIT Press

Rosenau, J. N. (1997). *Along the Domestic-Foreign Frontiers: Exploring Governance in a Turbulent World.* Cambridge: Cambridge University Press.

Ruggie, J. G. (1998). The Social Constructivist Challenge. *International Organization*, 53(4), 856-879.

Saurin, J. (2001). Global Environmental Crisis as the 'Disaster Triumphant': The Private Capture of Public Goods. *Environmental Politics*, 10(4), 63-84.

Singer, P. (2004). T*he President of Good & Evil: Questioning the Ethics of George W. Bush.* New York: Dutton.

Stadelmann, M., Michaelowa, A., Butzengeiger-Geyer, S. and Köhler, M. (2011). *Universal Metrics to Compare the Effectiveness of Climate Change* Adaptation Projects. Retrieved from http://www.oecd.org/env/cc/48351229.pdf

Stern, N. (2007). *The Economics of Climate Change. The Stern Review.* Cambridge: Cambridge University Press.

Shue, H. (1992). The Unavoidability of Justice. In: A. Hurrel and B. Kingsbury (eds). *International Politics of the Environment: Actors Interests and Institutions* (pp. 373-397). Oxford: Oxford University Press.

Shue, H. (1999). Global Environment and International Inequality. *International Affairs*, 75(3), 531-545.

Tanner, T. and Allouche, J. (2011). Towards a New Political Economy of Climate Change and Development. *IDS Bulletin*, 42(3), 1-14.

Vanderheiden, S. (2008). *Atmospheric Justice: A Political Theory of Climate Change*. Oxford: Oxford University Press.

Vanderheiden, S. (2011). Globalizing Responsibility for Climate Change. *Ethics and International Affairs*, 25(1), 66-84.

Volger, J. (1995). *The Global Commons: A Regime Analysis*. Chichester, UK: Wiley.

2

Perspectives of Global Environmental Change in the Anthropocene

ÚRSULA OSWALD SPRING
NATIONAL AUTONOMOUS UNIVERSITY OF MEXICO

The world faces economic crises, population growth, climate change, water scarcity and pollution, food crises, soil depletion, erosion and desertification, urbanisation with slum development, rural–urban and international migration, physical and structural violence, gender, race and ethnic discrimination, youth unemployment and an increasing loss of ecosystem services. The interaction of these multiple crises may result in extreme outcomes, especially for the vulnerable people living in risky places, and may reduce human, gender and environmental security.

This chapter addresses the complex interrelations and feedbacks between the environment system and the human system. It also explores the potential of multilateral negotiations among governments, organised society, and business community on long-term sustainable development goals.

Background

Climate change is a long-term process that acts in a context of climate variability in the short term, and with many influences on environment and humankind. It takes place at the regional and global scale. Historically, climate variability existed before and was produced by natural events, like volcanic eruptions and sun activity. Both water and carbon cycles together with other external parameters for the planet – position and activity of the sun

– have changed atmospheric conditions. Nevertheless, climate change is currently associated with human impacts on Earth (IPCC, 2013; 2014a; 2014b).

With climate change, temperature in the troposphere, over land and in the sea rises; water vapour increases; sea ice, glacier and permafrost lose volume; oceans maintain heat and energy, and sea level rise occurs due to the expansion of water and the melting glaciers. Linked to the interaction of these natural and human factors, extreme weather events (cyclones, droughts, landslides) occur more frequently and with stronger effects on many regions (IPCC, 2012).

Global environmental change is wider than climate change. The term refers to the transformation produced by human beings in the ecosphere and affecting the hydrosphere (the combined mass of water found above, on and under the surface of the planet), the atmosphere (the layer of gases surrounding the surface), the biosphere (the global ecological system where all living beings exist), the lithosphere (the outer layer of the earth) and the pedosphere (referring to the soil) (Brauch et al., 2008; 2009; 2011).

Changes in the natural system are the result of modifications in agricultural production, of rapid urbanisation processes, and of population growth—the human population tripled during the last century, but water consumption increased six-fold (Oswald, 2011). Furthermore, unsustainable productive processes are polluting natural resources and creating health threats for human beings, as well as endangering ecosystems (Elliott, 2011). Energy, transportation and production sectors have polluted heavily due to their use of fossil fuels (IEA, 2014). In addition, land-use change and deforestation are reducing the capture of carbon dioxide (CO_2) (IPCC, 2014a; 2014b). Hence, the emissions from greenhouse gases (GHG) have increased exponentially (IPCC, 2013).

In addition, a globalised financial system, unequal credit access, current patterns of consumption and production and uneven access to resources are also contributing to environmental change. Irrational behaviour has also produced poverty, hunger and inequality among regions and social groups (Wilkinson and Pickett, 2009).

In Earth and human history, gradually drastic changes have occurred since the industrial revolution (1780–1870). Crutzen (2002) links these changes with the concept of *Anthropocene*, which relates to environmental changes induced and produced predominantly by human interventions. Such changes have occurred especially during the last five decades due to the intensive use of fossil energy; the rapid increase in GHG emissions into the atmosphere; the pollution, warming and acidification of the seas; massive land-use changes; and an accelerated process of urbanisation. The Anthropocene represents a new geological epoch that is changing the history of Earth. Bond et al. (1997) defined it as 'the most recent manifestation of a pervasive millennial-scale climate cycle operating independently of the glacial-interglacial climate state'. This concept is useful for understanding the transformative negative effects of human activity on the global planet, its ecosystem services, and humankind itself. Nevertheless, human agency also has potential for positive change.

It is in this sense that the PEISOR model has been developed (Brauch and Oswald, 2009: 9). Based on the OECD's Pressure-State-Response model (2001), and by analysing the interaction among natural and human systems, this model combines five stages:

P: *pressure* refers to four natural drivers (climate change, water, soil and biodiversity), which interact with four social drivers (population growth, rural and urban systems and socioeconomic processes). The complex interaction and feedbacks cause environmental change.

E: *effects* of the interaction, where degradation and scarcity of natural resources produce stress, reinforced by urbanisation, productive processes, green revolution and population growth.

I: *impacts* of human-induced and climate-related natural hazards (storms, floods, landslides, droughts, forest fires, heat waves, cool spills), geophysical extreme events (earthquakes, tsunamis, volcanic eruptions) and technological or human-induced disasters (accidents, terrorism).

SO: *societal outcomes* represent the social response, where individual and community responses are analysed to overcome poverty, marginalisation and lack of education that often produce a survival dilemma: to stay at home, suffer and eventually die; to migrate and confront in the new place uncertainty of shelter, food, insecurity and labour; or to protest and fight for survival conditions at home. Crises, migration and conflicts may produce massive societal responses such as rapid urbanisation with slum development, violent outbreaks and internal crises, or conflict avoidance and peaceful resolutions of controversies, which enable negotiation processes for policy changes and institutional building.

R: *response* at local, regional, national or international level, where political processes involve the state, society and the business community to cope with global environmental change, reduce environmental stress, adapt to adverse conditions, develop resilience and build institutions, where traditional and scientific knowledge may help to restore a new equilibrium among socio-economic and environmental conditions.

The feedbacks among these different stages help to reduce the pressure and can promote disaster risks reduction (DRR) and disaster risk management (DRM), stimulate development processes and improve the global and local socio-economic, institutional and political contexts. Nevertheless, there are factors that often interact in unpredictable, non-linear and chaotic ways; challenging the society and the environment with possibly irreversible tipping points (Lenton et al., 2008). To overcome them, political negotiation processes may be needed to reduce environmental and social stress, reinforce adaptation and create institutions that can strengthen resilience from the bottom up with the support of sensitive top-down policies.

Pressure and actions on natural and social systems

Humankind uses the equivalent of 1.6 planet earths to provide the resources we need for consumption and to absorb our waste, and if we continue with the same pattern, in 2030 we will need two planets (Global Footprint Network, 2015). The extinction rate of species – compared with the pre-fossil age – is today 1,000 times higher and if humankind continues with the present unsustainable system of production and consumption it will be 10,000 times greater on average; affecting amphibians and birds, collapsing fisheries, diminishing forest cover, depleting fresh water systems (MA, 2005) and increasing GHG emissions. All these will increase the effects of environmental change. Moreover, scientists have warned that the earth will enter into the sixth largest extinction event – the first caused by human activities. Eighty per cent of CO_2 in the atmosphere now comes from energy used in transportation and industrial, economic and consumer activity; the rest is related to deforestation (IPCC, 2013) and destruction of ecosystems.

Two key indicators of a changing climate are temperature and sea level rise with changes in precipitation. According to the IPCC (2013), the global average air temperature over land and ocean surface has warmed by 0.85°C during 1880–2012. During 1901–2010, the global mean sea level rose 0.19 metres, with an increase from 1901 of 1.7 mm/year to 3.2 mm/year between 1993 and 2010, and precipitation changes impact regionally with extreme weather events.

With the increase of wellbeing and the consolidation of the middle classes in emerging countries, people moved up the food ladder – though eating meat is inefficient in terms of feeding everybody on Earth. However, grains have been diverted to industry and biofuels used or transportation. For example, 40 per cent of corn in the US is used for ethanol production (Foley, 2013). But this approach is unsustainable. One litre ethanol requires 2.37 kg of corn and between 1,200 to 3,600 litres of water; burns 500 g of coal and causes erosion of 15 to 25 kg soil. Despite this, the US subsidises this industry.

It will not be possible to promote efficient mitigation and adaptation actions without the involvement of exposed people, transparent support by governments and investment by the business community. Regional and local dual vulnerabilities may increase threats, and a collaborative interplay from bottom up and top down can reduce risks, especially when they are reinforced by international, national and local knowledge, global projections and multilateral and bilateral support.

The complex interrelationships between natural and human systems with feedbacks in the political and social arena – characterised by national and local contexts – show the mainstreaming of social vulnerability and its links to environmental change. Policy decisions affect the whole of the society and are tied to negotiation processes. Indeed, the history of high civilisations is instructive regarding environmental deterioration and the management of socio-political conflict.

The year 2015 has been regarded as the hottest year in history since systematic measurements were begun (WMO, 2015). Extreme hazards have increased worldwide because of global environmental change, with higher death rates and more affected people in the South and elevated economic damage in the North. Asia is the most exposed continent and its dual vulnerability (environmental and social – see Oswald et al., 2014) increases both the cost of disasters and human losses (EMDAT, 2015).

In 2008, food price hikes increased hunger worldwide. Between 800,000 and a billion people currently suffer from hunger (UNGA, 2015). Forty-four per cent of the world's population depends directly on ecosystem services for rain-fed agriculture and in 2014 two billion people were affected by flooding (EMDAT, 2015). The World Meteorological Organisation (WMO) and the International Organisation for Migration (IOM) estimate that, as a consequence of extreme climate events, environmentally induced migration will increase substantially. The presence of disease is likely to rise. Malaria, dengue, chikungunya and other tropical diseases are increasing with the higher temperature and spreading to higher altitudes, affecting people without adequate defences in their immune system (WHO, 2014).

Climate change effects are not gender neutral. Women are highly exposed to disasters caused by natural events (Ariyabandu and Fonseka, 2009). For example, Anttila-Hughes and Hsiang (2013) have claimed that post-typhoon economic deaths account for 13 per cent of the national infant mortality rate in the Philippines and that baby girls die 15 times more frequently, while for baby boys no increase in mortality rate was found. The long-term effects of this type of natural disaster add further to poverty.

Responses to global environmental change

The IPCC (2014a: 8) concluded that 'adaptation is becoming embedded in planning processes, with more limited implementation of responses. Engineered and technological options are commonly implemented adaptive responses, often integrated within existing programs such as disaster risk management and water management. There is increasing recognition of the value of social, institutional, and ecosystem-based measures and of the extent of constraints to adaptation.' Nevertheless, adaptation is often restricted to impacts, vulnerability and adaptation planning and preventive actions.

Global policy must limit the temperature increase to below 2°C above pre-industrial levels by the end of the century. This signifies a gradual process of decarbonising the economy, accompanied by a shift from fossil fuels to renewable energy (Ren21, 2015), the promotion of energy efficiency and the restoration of destroyed ecosystems. These actions should imply the dematerialisation of production, recycling of waste and adjustments to the existing model of civilisation and consumerism.

Global proactive policies of mitigation (Stern, 2006) can change the direction towards a sustainable transition (Grin et al., 2010), which may prevent an increasing number of disasters. Additionally, adaptive processes, precautionary learning, and resilience in communities exposed to environmental change allow developing capabilities needed to effectively protect people from future climate events. Restoration of coastal ecosystems, reforestation of slopes, land and environmental management, watershed sustainability and water protection support both mitigation and adaptation. Green agriculture and restoration of ecosystem services will not only improve water supply and air quality but also reduce the risks of disasters.

The IPCC (2014a; 2014b) explains that interaction in adaptation, mitigation and sustainable development occurs both within and across regions and scales, often in the context of multiple stressors. Some options to respond to climate change can imply risks of other environmental and social costs, have adverse distributional effects and draw resources away from development priorities such as poverty eradication.

To protect nature, people and their productive activities, an interdisciplinary cooperation among different epistemic communities is crucial. With this purpose in mind, climate change scientists have elaborated models for long-term climate policy and short-term prevention and early warning (IPCC, 2013; Dai, 2011). DRR and DRM communities' support for preventive and post-disasters activities is also noticeable (McBean and Ajibade, 2009; McBean, 2012), as well as socio-economic and cultural bases for resilience building (O'Brien et al., 2010; IPCC, 2014a; 2014b; World Bank, 2014). Moreover, there is a real chance of reducing risks to the most vulnerable people if a gender perspective is included (Ariyabandu and Fonseka, 2009; Fordham et al., 2011). Further, the potential interaction of these views can facilitate the adaptation to new and unknown risks (Beck, 2009; 2011). Simultaneously, it may also reduce damage to human lives and property and may help affected populations become more resilient and reduce their environmental and social vulnerability (Berkes, 2007).

From negotiations to extraordinary multilateral policy measures

The limited successes of post-Kyoto are raising these questions: who is managing the human securitisation process (Wæver, 1997) and under what conditions? The next question is which are the obstacles to overcome? Politics change radically when there is a shift from 'usual' political issues to a 'matter of security' of 'outmost importance'. There is consequently the need to develop an argument that goes beyond moral or ethical grounds, one that explores a combination of three securities, human, gender and environmental – or HUGE security (Oswald, 2009). This concept can be used as an analytic tool for analysis or policy guidance for proactive action.   By linking the PEISOR model with the HUGE security perspective, we suggest a broadening of the scope of conceptual, theoretical and empirical research on the climate–security nexus.

The policies of present business-as-usual management may produce a dangerous global change with an increasing number of catastrophes and irreversible tipping points. The HUGE security approach may have the potential to prompt multilateral negotiations among governments, organised society and the business community to achieve long-term sustainable development goals. These goals must offer even the most vulnerable livelihood and wellbeing, together with a systematic recovery of ecosystems and the services provided by such systems in relation to fresh water and the ocean. The challenge is to alter the model of concentrated global power based on multinational enterprises and military control in such a way as to enable long-term transformation towards sustainable transitions. This process requires a different political arena, without the dominance of any existing superpower and its control over people and resources. It is also necessary to change the Bretton Woods agreements and democratise the World Bank, International Monetary Fund and World Trade Organisation. Building new institutions from the bottom up and threatening disasters may achieve global sociopolitical contracts for a decarbonised and a dematerialised world with the potential to improve social equity and solidarity, which goes far beyond the voluntary agreements reached in the COP 21 in Paris 2015.

Observations

The HUGE security approach does not focus on security from military or political points of view, where the reference object is the state and the values at risks are sovereignty and territorial integrity. In the traditional understanding, the threats are related to other states, to terrorism and to sub-state actors or guerrillas. By focusing simultaneously on human, gender and environmental security, the reference shifts towards human beings, gender relations and natural, urban and rural ecosystems. The values at risk are the survival of humankind and nature, with equity, equality, identity, cultural diversity and sustainability in danger. Threats come from people themselves. They are also victims of this irrational behaviour. Changes must be made towards a new civilisation model, confirmed by a diverse, sustainable and global world where solidarity governs (Brauch et al., 2011). Who are the actors that can initiate and implement such a change? No social movement is doing it. The HUGE security approach could assist in analysing the best way to reach sustainable development goals, policies and strategies, out of which common but also differentiated responsibilities (CBDR) may offer ways forward.

Both concepts, HUGE and CBDR, are grounded in the Charter of the United Nations and the Universal Declaration of Human Rights, and both may be used to critically review existing international human rights conventions. In addition, the seventeen Sustainable Development Goals (SDGs) and their targets are initial roadmaps to challenge the present occidental global economic systems and the exacerbating injustice – since social equality, gender equity and sustainability are key elements.

The SDGs include small and smart economics, food sovereignty, fair international trade, alternative tax policies, and other bottom-up efforts, where private aid is scrutinised and transparency and accountability promoted by people. HUGE security also coincides theoretically with the 5 Ps – people, planet, prosperity, peace and partnership – with special focus on the reduction of dual vulnerability.

The required changes in the political arena imply transforming the model of governance. Participative governance (In't Veld, 2012) is needed, in which changing global arenas facilitate sustainable policies for water, air, climate, soil, food, biodiversity and energy. This requires negotiated agendas at local, national, regional and global levels, enabling policies to restore destroyed ecosystem services and overcome extreme poverty, hunger, illiteracy, diseases and disasters. Trained political actors and a critical participative civil society may be able to promote activities to achieve the common goals of a sustainable, equal and peaceful society in the 21st century.

References

Anttila-Hughes, J. K. and Hsiang, S. M. (2013). *Destruction, Disinvestment, and Death: Economic and Human Losses Following Environmental Disaster.* Retrieved from http://ssrn.com/abstract=2220501

Ariyabandu, M. M. and Fonseka, D. (2009). Do Disasters Discriminate? A Human Security Analysis of the Impact of the Tsunami in India, Sri Lanka and of the Kashmir Earthquake in Pakistan. In: H. G. Brauch, N. C. Behera, P. Kameri-Mbote, J. Grin, U. Oswald Spring, B. Chourou, C. Mesjasz and H. Krummenacher (eds). *Facing Global Environmental Change. Environmental, Human, Energy, Food, Health and Water Security Concepts* (pp. 1223-1236). Berlin: Springer.

Beck, U. (2009). *World at Risk.* Cambridge: Polity Press.

Beck, U. (2011). Living in and Coping with World Risk Society. In: H. G. Brauch, U. Oswald Spring, C. Mesjasz, J. Grin, P. Kameri-Mbote, B. Chourou, P. Dunay, and J. Birkmann (eds). *Coping with Global Environmental Change, Disasters and Security: Threats, Challenges, Vulnerabilities and Risks* (pp. 11–16). Berlin: Springer.

Berkes, F. (2007). Understanding Uncertainty and Reducing Vulnerability: Lessons from Resilience Thinking. *Natural Hazards*, 41, 283-295.

Bond, G., Showers, W., Cheseby, M., Lotti, R., Almasi, P., de Menocal, P., Priore, P., Cullen, H., Hajdas, I. and Bonani, G. (1997). A Pervasive Millennial-Scale Cycle in North Atlantic Holocene and Glacial Climates. *Science*, 278(5341), 1257–1266.

Brauch, H. G., Oswald Spring, U., Mesjasz, C., Grin, J., Dunay, P., Behera, N. C., Chourou, B., Kameri-Mbote, P. and Liotta, P. H. (eds) (2008). *Globalization and Environmental Challenges: Reconceptualizing Security in the 21st Century*. Berlin: Springer.

Brauch, H. G., Behera, N. C., Kameri-Mbote, P., Grin, J., Oswald Spring, U., Chourou, B., Mesjasz, C. and Krummenacher, H. (eds) (2009). *Facing Global Environmental Change. Environmental, Human, Energy, Food, Health and Water Security Concepts*. Berlin: Springer.

Brauch, H. G. and Oswald Spring, U. (2009). *Securitizing the Ground, Grounding the Security*. UNCCD Issue Paper No. 2. Bonn: UNCCD and Government of Spain.

Brauch, H. G., Oswald Spring, U., Mesjasz, C., Grin, J., Kameri-Mbote, P., Chourou, B., Dunay, P. and Birkmann, J. (eds) (2011). *Coping with Global Environmental Change, Disasters and Security: Threats, Challenges, Vulnerabilities and Risks*. Berlin: Springer.

Crutzen, P. J. (2002). Geology of Mankind. *Nature*, 415(6867), 23.

Dai, A. (2011). Drought under Global Warming: A Review. *Wiley Interdisciplinary Reviews*: Climate Change, 2(1): 45-65.

Elliott, M. L. (2011). First Report of Fusarium Wilt Caused by Fusarium oxysporum f. sp. palmarum on Canary Island Date Palm in Florida. *Plant Disease*, 95(3), 356.

EMDAT (2015). *Global Assessment Report on Disaster Risk Reduction*. Retrieved from http://www.preventionweb.net/english/hyogo/gar/2015/en/gar-pdf/GAR2015_EN.pdf.

Foley, J. (2013, March 5). It's Time to Rethink America's Corn System. *Scientific American*. Retrieved from http://www.scientificamerican.com/article/time-to-rethink-corn/

Fordham, M. and Gupta, S., with Akerkar, S. and Scharf, M. (2011). *Leading Resilient Development: Grassroots Women's Priorities, Practices and Innovations*. New York: Grassroots Organizations Operating Together in Sisterhood and the UN Development Programme.

Global Footprint Network (2015). *World Footprint: Do We Fit on the Planet?* Retrieved from http://www.footprintnetwork.org/ar/index.php/GFN/page/world_footprint/

Grin, J., Rotmans, J. and Schot, J. (2010). T*ransitions to Sustainable Development. New Directions in the Study of Long Term Transformative Change.* London: Routledge.

IEA (2014). *World Energy Investment Outlook.* Retrieved from www.iea.org/publications/freepublications/publication/weio2014.pdf

In't Veld, R. J. (2012). *Transgovernance. The Quest for Governance of Sustainable Development.* Potsdam: IASS.

IPCC (2012). Managing the Risks of Extreme Events and Disasters to Advance Climate Change Adaptation. A Special Report of Working Groups I and II of the Intergovernmental Panel on Climate Change. Cambridge: Cambridge University Press.

IPCC (2013). C*limate Change 2013: The Physical Science Basis. Contribution of Working Group I to the Fifth Assessment Report of the Intergovernmental Panel on Climate Change.* Cambridge: Cambridge University Press.

IPCC (2014a). *Climate Change 2014: Impacts, Adaptation, and Vulnerability. Part A: Global and Sectoral Aspects. Contribution of Working Group II to the Fifth Assessment Report of the Intergovernmental Panel on Climate Change.* Cambridge: Cambridge University Press.

IPCC (2014b). *Climate Change 2014. Mitigation of Climate Change. Working Group III Contribution to the Fifth Assessment Report of the Intergovernmental Panel on Climate Change.* Cambridge: Cambridge University Press.

Lenton, T. M., Held, H., Kriegler, E., Hall, J. W., Lucht, W., Rahmstorf, S. and Schellnhuber, H. J. (2008). *Tipping Elements in the Earth's Climate System*, PNAS 105(6), 1786–1793. 12 February. Retrieved from http://www.pnas.org/content/105/6/1786.full.pdf

MA [Millennium Ecosystem Assessment] (2005). *Ecosystems and Human Wellbeing*: Desertification Synthesis. Washington, DC: Island Press.

McBean, G. A. (2012). Integrating Disaster Risk Reduction towards Sustainable Development. *Current Opinion in Environmental Sustainability*, 4, 122–127.

McBean, G. and Ajibade I. (2009). Climate Change, Related Hazards and Human Settlements. *Current Opinion in Environmental Sustainability*, 1(2), 179–186.

O'Brien, K., Lera St. Clair, A. and Kristoffersen, B. (eds) (2010). *Climate Change, Ethics and Human Security*. Cambridge: Cambridge University Press.

OECD (2001). *OECD Environmental Indicators. Development Measurement and Use*. Geneva: OECD.

Oswald Spring, U. (2009). A HUGE Gender Security Approach: Towards Human, Gender, and Environmental Security. In: H. G. Brauch, N. C. Behera, P. Kameri-Mbote, J. Grin, U. Oswald Spring, B. Chourou, C. Mesjasz and H. Krummenacher (eds). *Facing Global Environmental Change. Environmental, Human, Energy, Food, Health and Water Security Concepts* (pp. 1165-1190). Berlin: Springer.

Oswald Spring, U. (2011). *Water Resources in Mexico. Scarcity, Degradation, Stress, Conflicts, Management, and Policy.* Berlin, Heidelberg and New York: Springer.

Oswald Spring, U., Serrano Oswald, S. E., Estrada Álvarez, A., Flores Palacios, F., Ríos Everardo, M., Brauch, H. G., Ruiz Pantoja, T. E., Lemus Ramírez, C., Estrada Villanueva, A. and Cruz Rivera, M. T. M. (2014). *Vulnerabilidad Social y Género entre Migrantes Ambientales.* Cuernavaca: CRIM-DGAPA-UNAM.

Ren21 (2015). *Renewables 2015. Global Status Report*. Retrieved from http://www.ren21. net/wp-content/uploads/2015/07/REN12-GSR2015_Onlinebook_low1.pdf

Stern, N. (2006). *The Economics of Climate Change: The Stern Review.* Cambridge: Cambridge University Press.

UNGA [UN General Assembly] (2015, August 5). Right to Food. A/70/287. New York: UN. Retrieved from http://www.refworld.org/docid/55f291324.html

Wæver, O. (1997). *Concepts of Security*. Copenhagen: Department of Political Science.

WHO [World Health Organization] (2014). *Dengue and Malaria Impacting Socioeconomic Growth*. WHO, SEAR/PR 1570. Retrieved from http://www.searo.who.int/mediacentre/releases/2014/pr1570/en/

Wilkinson, R. and Pickett, K. (2009). *The Spirit Level: Why More Equal Societies Almost Always Do Better*. London: Allen Lane.

WMO [World Meteorological Organization] (2015, November 25). *WMO: 2015 likely to be Warmest on Record, 2011-2015 Warmest Five Year Period*. Press Release No. 13. Retrieved from https://www.wmo.int/media/content/wmo-2015-likely be-warmest-record-2011-2015-warmest-five-year-period

World Bank (2014). *Risk and Opportunity, World Development Report*. Washington, DC: World Bank.

3

Environment and International Politics: Linking Humanity and Nature

SIMON DALBY
WILFRID LAURIER UNIVERSITY, CANADA

Environment is now a key component of international relations and, given the rising attention climate change receives in particular (Welzer, 2012), a matter that now has high priority in diplomatic circles. With states in danger of disappearing below rising seas and major disruptions to water supplies and food systems projected for future decades if steps to curb greenhouse gas emissions are not taken soon, environmental matters have become central to contemporary international politics, and to their academic study (Webersik, 2010). Environment emerged after the Cold War as a priority matter for scholarly analysis because scholars are concerned with matters of pollution, conservation and resources; but also because there are interesting analytical puzzles surrounding how the international system deals with them and the changes resulting from the introduction of new modes of governance, institutions, agencies, knowledge and norms.

How all these aspects might be studied as international relations is also not a simple matter, but given their importance they have increasingly impinged on scholarship. Likewise, the rise of International Relations as an Anglo-American 'discipline' – a matter more closely related to the rise of industrial powers than usually acknowledged (Ashworth, 2014) – has shaped the kinds of questions asked about environment and the assumptions about how environmental politics is to be included in the field. Frequently, this has led to technical issues of regime design, compliance and funding mechanisms being focused on matters of justice or perspectives from marginal places. A partial response to this were interesting case studies on social movements and how their norms have impinged on the formal deliberations of states and intergovernmental agencies (Lipschutz and Mayer, 1996), a matter of governance more widely understood than the narrower concerns of formal state government (Young, 1994).

Critics, who argue that environmental management arrangements that focus on technical matters frequently occlude complicated processes of global injustice and the displacement of marginal populations in the global polity, have challenged this narrow analytic focus and raised larger questions of global power, justice and conflict (Sachs and Santarius, 2007). The result is an intense series of academic debates in International Relations and cognate fields about what to study, for whom and with what policy implications for governance broadly conceived. Only sometimes do the traditional core themes of International Relations concerned with war, peace and security impinge directly on the environmental discussions.

'The environment'

Many of the themes now frequently included under the rubric 'environment' are not necessarily understood in these terms by those who are affected by atmospheric change, water purity concerns, species loss, industrial pollution, land appropriation, deforestation and numerous other practices. The term 'environment' itself has also recently encompassed longstanding human debates about the role of nature in shaping the human fate and how humanity has in turn transformed natural conditions (Marsh, 1864; Glacken, 1967; Robin et al., 2013). Using environment as a term often lumps concerns with industrial pollution, technical fixes to production systems criticised for causing consumption (Dauvergne, 2008), and fears on the part of many in the Global North that population growth will overwhelm agricultural productivity, leading to famine and social disaster (Robertson, 2012).

There has been a long-standing suspicion, by at least some in the Global South, that the formulation of environment is one that is used to control Southern peoples; clearly, in many cases environmental measures are used as a justification for undertaking development projects that cause displacements and suffering for poor people all in the name of universal causes (Miller, 1995). Traditional modes of managing forests, limiting hunting, and other communal arrangements are frequently not a good fit for state-based governance structures, the entities that are usually the subject of international agreements. The extensive use of the term sustainable development, now codified in the recently adopted overarching 2015 United Nations (UN) Sustainable Development Goals (SDGs), has long been a compromise argument ostensibly dealing with environmental protection while simultaneously offering aid and compensation to Southern states for what is sometimes understood as forgone development opportunities. Likewise, this rubric encompasses all sorts of technical innovations, matters of ecological modernisation that supposedly allow industrial processes to proceed without pollution. These discussions have been key to the dominant concern in recent environmental matters, the question of climate change, and how to tackle what is now understood as a global problem (Bulkeley et al., 2014).

Most recently, as the sheer scale of human activities gradually dawns on policymakers and the interconnections between various Earth system processes become clearer, both in new historical research (Davis, 2001; McNeill 2000; Hornborg, et al., 2007) and in scientific assessments of global change (Ellis, 2011; Steffen et al., 2004; United Nations Environment Program, 2012); the environment discussions have increasingly focused on how the rich and powerful parts of humanity will shape the future configuration of the planet. Clearly there are numerous failures of governance in trying to tackle the interconnected problems of what is increasingly called the age of the Anthropocene (Galaz, 2014). Will rapid climate change lead to deliberate attempts to change the atmosphere to slow or counteract global warming as in geoengineering, or will the powerful states and corporations act quickly to preclude the necessity of such drastic, and potentially conflict causing measures? While it may be premature to call current circumstances 'the age of ecology' (Radkau, 2014), such considerations are increasingly shaping matters of global politics.

Science and politics

Prior to the 1960s there were precursors to the idea of a single global entity that might be regulated and managed; such things as conventions on migratory birds, like the one signed by the United States (US) and the United Kingdom (UK), on behalf of Canada in 1918, did attempt to grapple with what is now understood as the international dimensions of nature conservation. But it is only in the second half of the twentieth century that these became a focus for widespread attention by academics and policymakers (see Brown, 1954; Thomas, 1956). This has been driven by a combination of rapid economic growth, political pressure from domestic environmental constituencies worried about pollution, population, parks and nature protection, and growing international environmental organisations epitomised by the rise of Greenpeace in the 1970s, as well as crucial innovations in science that have focused attention on issues that require international cooperation to address, perhaps mostly pointedly, the depletion of stratospheric ozone.

The rise of concern about what became known as a global environment is also in part a spin-off of Cold War concerns with geophysics. The international geophysics year (in fact 18 months) in 1957/8 was driven in part by efforts and scientific cooperation across the Cold War divide, but also by military concerns about dominating and controlling atmospheric spaces. (What has become the iconic graph of our times, the so-called Keeling curve of rising carbon dioxide concentrations in the atmosphere measured atop a mountain in Hawai'i, has its origins in the international geophysics year.) The fallout from nuclear weapons tests carried around the earth by winds made it clear that the global atmosphere was one interconnected system. These concerns lead to the partial nuclear test ban treaty in the early 1960s, a treaty that was simultaneously an attempt to constrain the arms race between the superpowers, and one that was also the first global atmospheric environmental treaty (Soroos, 1997).

As Edwards (2010) makes clear, these scientific endeavours, and in particular the emergence of meteorology and weather forecasting, and the subsequent invention of weather satellites have been crucial parts of the rise of a global sensibility as the backdrop for human activities. While global trade and television may have knitted the world together in the processes we have now, after the Cold War, come to call globalisation, some of the key factors have been environmental sciences and the practical spin-off of relatively reliable weather forecasts. Likewise, damage done to the stratospheric ozone layer from high altitude nuclear tests raised further concerns about the upper atmosphere in the 1950s. These issues of ozone depletion subsequently came to a head in the 1980s when stratospheric depletion over the Antarctic caused a flurry of scientific investigations that confirmed that chlorofluorocarbons (CFC) were scavenging ozone in the upper atmosphere and threatening terrestrial life due to the increased penetration of solar Ultraviolet B (UVB) radiation. The resultant scientific debate fed directly into negotiations to eliminate CFC production globally and the Montreal Protocol of 1987 and subsequent extensions of these arrangements to constrain other ozone depleting substances (Litfin, 1994).

Science also drove rising concerns about pollution in the 1950s, although the huge death toll from coal fire-generated smog in London, in particular, did not need much scientific analysis to explain what happened or prompt the passing of clean air legislation in the UK. In the US, Rachel Carson's (1962) book *The Silent Spring*, with its analysis of the indirect damage that widespread pesticide use had caused to ecosystems and bird populations, drew attention to the unintended effects of the chemical industry and brought intense pressure to bear to produce regulation of chemical pollution. Automobile smog in Los Angeles emphasised the environmental hazards of industrial production and suburbanisation, and connected up with longstanding issues of urban reform and city planning (Howard, 1898).

Concerns about resource depletion and shortage of key commodities have long been at the heart of geopolitical concerns (Le Billon, 2012). General fears of resource depletion have long preoccupied state governments; in the late eighteenth century it was fears of wood shortage and inadequate forestry management. Thomas Malthus (1970) feared that the population would grow faster than its ability to feed itself, and his famous essay has shaped many of the predominant modern narratives that specify scarcity as the human condition, even as the extraordinary productivity of industrial systems repeatedly belies the assumption. Likewise, in the 1960s Paul Ehrlich (1968) linked population concerns drawn from ecological studies of wildlife to the larger angst over pollution and resource depletion in the US with his bestseller 'The Population Bomb'. Early attempts to model the Earth system included such resource exhaustion patterns and produced a hugely popular report on 'The Limits to Growth' (Meadows et al., 1974). Richard Falk (1971) explicitly linked population, inadequate food production and fears of nuclear apocalypse into a discussion of international relations and global political reform.

These themes all interconnected in the 1960s in the US and the UK, in particular, and generated what is now known as the environmental movement (O'Riordan, 1976). The Greenpeace innovation of taking non-violent direct action to the high seas and using television footage of the confrontations created political dramas that highlighted the politics of environment (Wapner, 1996). Simultaneously, the first pictures of Earth from the Apollo moon programme showed a fragile blue marble set against the dark backdrop of space confirming a sensibility of global interconnectedness. A flurry of environmental legislation was passed in the US at the end of the 1960s and in the early 1970s. The first 'Earth Day' happened in 1970.

Global environment

All this generated considerable political attention outside the US too, and the UN Conference on the Human Environment was convened in Stockholm in 1972. The unofficial background report to the conference was written by Barbara Ward and Rene Dubos (1972) and titled 'Only One Earth'. The NASA cover photograph of 'Earth rising' on the British Penguin edition encapsulated the key message of the report. The conference was boycotted by the Warsaw Pact countries and attended by only a few heads of state, but generated considerable attention – not least when Indira Gandhi insisted that poverty was the worst kind of pollution and that developed states should not use environmental arguments to stymie the development aspirations of poor and post-colonial states. Perhaps the most important legacy of this conference is that these matters were firmly placed on the international agenda and the, admittedly very poorly resourced, United Nations Environment Program (UNEP) was established to move deliberations on how the international community should respond ahead.

Subsequently, discussions on the international law of the sea and notions of the common heritage of mankind broadened concerns about environmental matters to cover the oceans as well as terrestrial, outer space and atmospheric matters (Vogler, 1995). Environmental change and geopolitics were once again interconnected in 1983, when research into the possibilities of a nuclear winter in the aftermath of a nuclear war between the superpowers suggested that prompt fatalities from nuclear detonations would be followed by a dramatic fall in global temperatures due to dust and smoke in the atmosphere (Turco et al., 1983). The ecological consequences from this rapid onset climate change might, it was argued, terminate civilisation. These discussions once again linked climate to the central concerns of international relations, and simultaneously made it clear that human activity was capable of changing the basic geophysics of the planetary atmosphere.

Nuclear winter concerns were supplemented by both the Chernobyl nuclear reactor meltdown in 1986 and the growing alarm about ozone depletion. Simultaneously, discussions of sustainable development were afoot leading to the publication of the World Commission on Environment and Development's *Our Common Future* in 1987. The 'Brundtland Report' as it is often called, after the Norwegian chair of the commission, set the stage for the huge UN Conference on Environment and Development in Rio de Janeiro in 1992, where the United Nations Framework Convention on Climate Change (UNFCCC) was launched. But critics were quick to point out that despite all the hype about saving the world and dealing with development issues, the rich and powerful states and corporations were primarily concerned with business as usual rather than dealing with poverty or new modes of economic activity that would make the future sustainable for marginal peoples and places (*The Ecologist*, 1993).

These rising concerns with environment, as the Cold War wound down, followed many of the dominant themes in International Relations scholarship at the time. These approaches continue to shape how many studies are formulated. Partly their impetus comes from international relations but it is important to emphasise that they are also shaped by how environmental issues are framed in domestic politics, and by larger political trends. The focus on international organisations and liberal political economy has shaped much of the discussion of international regimes and the importance of epistemic communities in facilitating agreements (Keohane and Nye, 1977).

International Relations' core concern has long been with warfare and the dangers of and how to prevent international conflict. In the aftermath of the Cold War conflict theorists looked at environmental conflict and the possibilities of resource conflicts as a source of warfare. Larger concerns with political economy have emphasised the importance of global inequities and the role of production and trade in shaping how pollution, land use and resource extractions play out across the globe.

Regimes

Garrett Hardin (1968) published a hugely influential article that suggested that many resource problems could be understood in terms of a 'tragedy of the commons', a misleading historical analogy that has generated numerous studies in environmental politics. Complaints from activists in the 1970s (Roberts, 1979) and subsequent careful work on resource systems by Elinor Ostrom (1990) – for which she eventually won the Nobel economics prize – make clear that commons systems have frequently had effective management systems, and that the enclosure and removal of traditional patterns by the expansion of extractive commercial arrangements frequently causes degradation. But the analogy is very suggestive in international relations, where oceans and atmosphere have no central authority to regulate activities and regime theorists frequently invoke Hardin's (1968) formulation in discussions of, as in Vogler's (1995) terms, the global commons.

This approach, focusing on the design of international agreements and the institutional innovations and norms that support them, draws on liberal international thinking and focuses on the collaborative possibilities in international matters that shape treaties and protocols to govern such matters of commons beyond the jurisdiction of states (Young, 1994). It has been extended to analysis of aid arrangements and also development assistance related to environmental management (Haas et al., 1993). Frequently, these are matters of trade restrictions, using mechanisms to prevent the transboundary movement of commodities, such as in the case of ivory as a way to remove financial incentives for killing animals. The Convention on the International Trade of Endangered Species is an exemplary case.

Crucial to the formation of many regimes are matters of technical knowledge and, related to this, technical standards for measuring and monitoring environmental matters (Haas, 1990). The construction of epistemic communities with shared scientific knowledge and agreed modes of specifying problems and crafting procedures and techniques are necessary prerequisites for international management of complicated problems, like ozone depletion. The case of ozone emphasises the point that environmental matters are often highly technical (Litfin, 1994). No one can actually see the depletion of stratospheric ozone; the complicated chemistry of ozone scavenging involves the breakdown of CFCs and reactions catalysed by ice particles in stratospheric clouds over Antarctica, and the measurement of ozone concentrations is a technical matter that requires complicated measuring devices. Hence, the importance of shared scientific understandings in drafting agreements and ensuring compliance. Despite the relevance of dealing with ozone depletion, it is worth emphasising the various industries, notably strawberry growers in California, that have claimed strategic exemption from such arrangements; a pattern of industrial evasion and delay in dealing with environmental regulations connected up with political campaigns of obfuscation and denial that has hampered governance efforts only most obviously in the case of climate change (Jacques, 2009).

Environmental security

In the context of the 1980s, with rising worries about ozone, fallout from Chernobyl, rapid deforestation and burning in Brazil, chemical disasters such as Bhopal, and nascent concerns about climate change causing weather disruptions and water shortages; it was not hard to point to the insecurities that environmental matters caused in many places. The World Commission on Environment and Development (1987) suggested that poor use of resources and rising scarcities might well lead to conflict and, as such, sustainable development was understood as a necessary prophylactic.

But as Thomas Homer-Dixon (1991) pointed out, the simple assumption in the policy discussion that this was the next major security threat after the Cold War dissipated needed serious scholarly analysis before it could be claimed as a security issue. The subsequent discussion linking security to environment suggested that environmental change frequently provided opportunities for cooperation rather than conflict, and that where environmental conflict did occur it was highly unlikely to lead to interstate warfare however much small-scale violence might be entailed (Homer-Dixon, 1999). Early on in this discussion, Daniel Deudney (1990) argued that the military was probably the last institution that should be used to deal with environmental matters given that it was singularly ill equipped for the practical tasks at hand.

All this has been revisited more recently in discussions of climate and the possible security implications of a rapidly changing world, work that suggests more radical rethinking of the role of modern states in the provision of security (Brauch et al., 2011; Dalby, 2009). The critical work on environmental conflict in the 1990s also pointed to the importance of understanding how the global political economy was driving environmental change (Suliman, 1999). Likewise, it has become increasingly clear that the processes of development are frequently very disruptive to rural communities and traditional ways of life; a matter now often understood in terms of slow violence around particular extractivist resource projects (Nixon, 2011). Quite who is insecure and where needs careful analysis in discussions of environmental security (Barnett, 2001).

Political economy

Industrial responses to environmental regulation followed a number of strategies in response to the rise of environmental concerns from the 1960s onwards. The most obvious innovations were a series of technical fixes to prevent pollution. Scrubbers and filters on effluent pipes were followed by more sophisticated processes that aimed to eliminate waste products by being much more efficient in the use of raw materials. Sophisticated permit systems that permitted cleaner producers to sell pollution allowances to less efficient producers with an overall industry wide cap on emissions were used to curtail acid rain in North America (Ellerman et al., 2000), a cap and trade market mechanism that is now being reinvented in attempts to deal with climate change. All these innovations, a matter of ecological modernisation (Mol, 2003), use sophisticated regulations to reduce the environmental burden of particular industries but do not address the overall expansion of the global economy nor the inequities in its system. As such, these fit with the overall political economy of neoliberalism, in which states facilitate capital accumulation as first priority and use market mechanisms to do so.

While attention to international regimes and the epistemic communities that link rules and regulation to environmental technical practices focuses on the structure and dynamics of international organisations and the finer points of state strategies in the bargaining processes that create regimes, there is a larger literature that draws from more critical work in political economy looking at production, accumulation and the role of wealth and power in shaping environmental politics (Elliott, 2004; Newell, 2012). The subtitle to Matthew Paterson's 2000 volume, *Understanding Global Environmental Politics*, is phrased 'Domination, Accumulation, Resistance' to emphasise the importance of the political economy in resource and environment issues.

None of this is surprising given the roots of many discussions of resources in Thomas Malthus' pessimistic perspective, nor is the importance of trade measures in the international regulation of environmental matters. But it does point to the fact that a focus solely on regimes and forms of knowledge is always in danger of losing sight of the economic factors that drive production decisions, land use allocations and the effectiveness of government regulation of pollution and resource use (Stevis and Assetto, 2001). Clearly, if the world is to find pathways to green futures, these issues of political economy will be crucial (Clapp and Dauvergne, 2005). All of these issues have become especially pressing in the discussions of climate change.

Climate change

Thinking about climate change in terms of a regime similar to the one that has successfully constrained ozone depletion (and in particular restricted greenhouse gas emissions) – while dealing with the pattern of demands from developing countries that developed ones provide financial compensation for climate adaptation and forgone development projects that might rely on carbon fuel systems – has had limited success despite the Kyoto Protocol negotiated in the 1990s. Much of the recent International Relations scholarship on climate change has looked at the finer points of negotiation, the strategies of bargaining at the Conferences of the Parties (COP) of the UNFCCC and more recently the alliances between various international 'clubs' (Nordhaus, 2015).

Research has also focused on the financial mechanisms that have been used as tools in attempts to deal with climate change, carbon offsets and clean development mechanisms and related market arrangements in the burgeoning cap and trade schemes around the world (Newell and Paterson, 2010). But the inadequacy of these arrangements have become a pressing issue in climate policy and International Relations scholars have been looking at more complex ideas of governance that spread well beyond the traditional arenas of interstate relations (Bulkeley et al., 2014). In the process, they challenge political scientists to think through what political incentives might move climate policy forward more efficaciously (Keohane, 2015). This not least because of the continuing discrepancies between Northern and Southern perspectives on what needs to be done and who is to blame for climate change (Chaturvedi and Doyle, 2015).

All this seems to be necessary because there is obviously a large gap between existing governance mechanisms and the tasks that need to be tackled in a rapidly changing world (Galaz, 2014). To do so may require that other aspects of international relations or global politics engage with the current global situation. Recently, Ken Conca (2015) has pointed out that within the UN system, environmental matters have largely been disconnected from efforts to think seriously about human rights and peace. Dealing with environment as a matter of better laws between states and better forms of development within them has ignored UN concerns with peace and with human rights. Perhaps, Conca (2015) suggests, tackling environment in terms of rights and peace making might lead to more useful advances in both policy and scholarly inquiry.

Anthropocene futures

While state sovereignty is a principle of world order that underpins the current system, it has long been clear that environmental matters are no respecter of frontiers (Leichenko and O'Brien, 2008). This insight is exemplified by the use of the term Anthropocene to emphasise that the rich and powerful parts of humanity are causing the sixth extinction event in the planet's history (Kolbert, 2014), while transforming numerous facets of the biosphere (Steffen et al., 2011). This formulation is now key to innovative thinking that transcends the intellectual strictures of the state system (Hamilton, Bonneuil and Gemenne, 2015).

Campaigns to tackle climate change are taking shape in many parts of the world, tied into protests against the depredations of mines, forest clearing and dam building, and other extractivist projects (Klein, 2014). These protest campaigns in turn are once again raising questions about the locus of authority in the global system and hence leading to further work by scholars on the role of social movements and global civil society in shaping international relations and a larger critical evaluation of the limits of traditional approaches to environmental politics (Death, 2014).

Failure to deal seriously with climate change, in particular, and the larger transformations of other Earth system elements already underway, in general (Steffen et al., 2015), is likely to lead to arguments for artificially modifying Earth system temperatures by such geoengineering projects as stratospheric aerosol injection (Burns and Strauss, 2013). While this thoroughly alarms critics of environmental modification, not least because of the potential of such projects to exacerbate international tensions (Hamilton, 2013), the future is likely to see such discussions rise in prominence in international relations unless policies for dealing with global change become much more effective soon.

Given the speed and scale of global transformations now in motion and the failures to integrate matters of ecology into larger concerns with peace, human rights and militarism (Amster, 2015), the old environmentalist question of 'who speaks for Earth?' is an ever more pressing issue for International Relations. Among the key new themes for the current generation of scholars are questions of how to end the fossil fuel era (Princen et al., 2015) and the urgent matter of facilitating transitions to much more sustainable patterns of life.

References

Amster, R. (2015). *Peace Ecology.* Boulder: Paradigm.

Ashworth, L. (2014). *A History of International Thought.* London: Routledge.

Barnett, J. (2001). *The Meaning of Environmental Security.* London: Zed.

Brauch, H. G., Oswald Spring, U., Kameri-Mbote, P., Mesjasz, C., Grin, J., Chourou, B., Dunay, P. and Birkmann, J. (eds) (2011). *Coping with Global Environmental Change, Disasters and Security Threats, Challenges, Vulnerabilities and Risks.* Berlin-Heidelberg – New York: Springer-Verlag.

Brown, H. (1954). *The Challenge of Man's Future.* New York: Viking.

Bulkeley, H., Andonova, L., Betsill, M. M., Compagnon, D., Hale, T., Hoffmann, M. J., Newell, P., Paterson, M., Roger, C. and VanDeveer, S. D. (2014). *Transnational Climate Change Governance.* Cambridge: Cambridge University Press.

Burns, W. C. G. and Strauss A. L. (eds) (2013). *Climate Change Geoengineering: Philosophical Perspectives: Legal Issues and Governance Frameworks.* Cambridge: Cambridge University Press.

Carson, R. (1962). *Silent Spring.* Boston: Houghton Mifflin.

Chaturvedi, S, and Doyle, T. (2015). C*limate Terror: A Critical Geopolitics of Climate Change.* New York: Palgrave Macmillan.

Clapp, J. and Dauvergne, P. (2005). *Paths to a Green World: The Political Economy of the Global Environment.* Cambridge, MA: MIT Press.

Conca, K. (2015). *An Unfinished Foundation: The United Nations and Global Environmental Governance.* New York: Oxford University Press.

Dalby, S. (2009). *Security and Environmental Change.* Cambridge: Polity.

Dauvergne, P. (2008). *The Shadows of Consumption: Consequences for the Global Environment.* Cambridge, MA: MIT Press.

Davis, M. (2001). *Late Victorian Holocausts: El Nino Famines and the Making of the Third World*. London: Verso.

Death, C. (ed.) (2014). *Critical Environmental Politics*. New York: Routledge.

Deudney, D. (1990). T*he Case Against Linking Environmental Degradation and National Security*. Millennium, 19(3), 461-476.

The Ecologist (ed.) (1993). *Whose Common Future? Reclaiming theCcommons*. London: Earthscan Publications.

Edwards, P. N. (2010). *A Vast Machine: Computer Models, Climate Data, and the Politics of Global Warming*. Cambridge, MA: MIT Press.

Ehrlich, P. (1968). *The Population Bomb*. New York: Ballantine.

Ellerman, A. D., Joskow, P. L., Schmalensee, R., Montero, J. P. and Bailey, E. M. (2000). *Markets for Clean Air: The US Acid Rain Program*. Cambridge: Cambridge University Press.

Elliot, L. (2004). *The Global Politics of the Environment*. New York: Palgrave Macmillan.

Ellis, E. C. (2011). *Anthropogenic Transformation of the Terrestrial Biosphere. Philosophical Transactions of the Royal Society A*, 369, 1010-1035.

Falk, R. A. (1971). *This Endangered Planet: Prospects and Proposals for Human Survival*. New York: Random House.

Galaz, V. (2014). *Global Environmental Governance, Technology and Politics: The Anthropocene Gap*. Cheltenham: Edward Elgar.

Glacken, C. (1967). *Traces on the Rhodian Shore*. Berkeley, CA: University of California Press.

Haas, P. M. (1990). Obtaining International Environmental Protection through Epistemic Communities. *Millennium*, 19(3), 347-364.

Haas, P. M., Keohane, R. and Levy, M. (1993). *Institutions for the Earth: Sources of effective international environmental protection.* Cambridge, MA: MIT Press.

Hamilton, C. (2013). E*arthmasters: The Dawn of the Age of Climate Engineering.* New Haven, CT: Yale University Press.

Hamilton, C., Bonneuil, C. and Gemenne, F. (eds) (2015). *The Anthropocene and the Global Environmental Crisis: Rethinking Modernity in a New Epoch.* Abingdon: Routledge.

Hardin, G. (1968). The Tragedy of the Commons. *Science*, 162, 1243-1248.

Homer-Dixon, T. (1991). On the Threshold: Environmental Changes as Causes of Acute Conflict. *International Security*, 16(2), 76-116.

Homer-Dixon, T. (1999). E*nvironment, Scarcity, and Violence.* Princeton, NJ: Princeton University Press.

Hornborg, A., McNeill, J. R. and Martinez-Alier, J. (Eds) (2007). R*ethinking Environmental History: World System and Global Environmental Change.* Lanham, MD: Altamira.

Howard, E. (1898). *To-morrow: A Peaceful Path to Real Refor*m. London: Swan Sonnenschein & Co.

Jacques, P. J. (2009). *Environmental Skepticism: Ecology, Power and Public Life.* Farnham: Ashgate.

Keohane, R. O. (2015). The Global Politics of Climate Change: Challenge for Political Science. *PS: Political Science & Politics*, 48(1), 19-26.

Keohane, R. O. and Nye Jr., J. S. (1977). *Power and Interdependence.* New York: Addison Wesley Longman.

Keohane, R. O. and Victor, D. G. (2011). The Regime Complex for Climate Change. *Perspectives on Politics*, 9(1), 7-23.

Klein, N. (2014). T*his Changes Everything.* Toronto: Knopf.

Kolbert, E. (2014). *The Sixth Extinction: An Unnatural History*. New York: Henry Holt.

Leichenko, R. L. and O'Brien, K. (2008). *Environmental Change and Globalization: Double Exposures*. Oxford: Oxford University Press.

Le Billon, P. (2012). *Wars of Plunder*. London: Hurst.

Litfin, K. (1994). *Ozone Discourses: Science and Politics in International Environmental Cooperation*. New York: Columbia University Press.

Lipschutz, R. and Mayer, J. (1996). *Global Civil Society and Global Environmental Change: The Politics of Nature from Place to Planet*. New York: State University of New York.

Malthus, T. (1798/1970). *An Essay on the Principle of Population*. Harmondsworth: Penguin.

Marsh, G. P. (1864/1965). *Man and Nature: Or, Physical Geography as Modified by Human Action*. Cambridge, MA: Harvard University/ Belknap Press.

McNeill, J. R. (2000). *Something New Under the Sun: An Environmental History of the Twentieth Century World*. New York: Norton.

Meadows, D. H., Meadows, D. L., Randers, J. and Behrens III, W. W. (1974). *The Limits to Growth*. London: Pan.

Miller, M. A. L. (1995). *The Third World in Global Environmental Politics*. Boulder, CO: Lynne Rienner.

Mol, A. (2003). *Globalization and Environmental Reform: The Ecological Modernization of the Global Economy*. Cambridge, MA: MIT Press.

Newell, P. (2012). *Globalization and the Environment: Capitalism, Ecology and Power.* Cambridge: Polity.

Newell, P. and Paterson, M. (2010). *Carbon Capitalism: Global Warming and the Transformation of the Global Economy*. Cambridge: Cambridge University Press.

Nixon, R. (2011). *Slow Violence and the Environmentalism of the Poor*. Cambridge, MA: Harvard University Press.

Nordhaus, W. (2015). Climate Clubs: Overcoming Free-riding in International Climate Policy. *American Economic Review 2015*, 105(4), 1339-1370.

O'Riordan, T. (1976). *Environmentalism*. London: Pion.

Ostrom, E. (1990). *Governing the Commons*. Cambridge, MA: Cambridge University Press.

Paterson, M. (2000). *Understanding Global Environmental Politics: Domination, Accumulation, Resistance*. New York: St Martin's Press.

Princen, T., Manno, J. P. and Martin, P. L. (eds) (2015). *Ending the Fossil Fuel Era*. Cambridge, MA: MIT Press.

Radkau, J. (2014). *The Age of Ecology: A Global History*. Cambridge: Polity.

Roberts, A. (1979). *The Self-Managing Environment. London*: Alison and Busby.

Robertson, T. (2012). *The Malthusian Moment: Global Population Growth and the Birth of American Environmentalism*. New Brunswick: Rutgers University Press.

Robin, L., Sörlin, S. and Warde, P. (2013). *The Future of Nature: Documents of Global Change*. New Haven, CT: Yale University Press.

Sachs, W. and Santarius, T. (eds) (2007). F*air Future: Resource Conflicts, Security and Global Justice*. London: Zed.

Soroos, M. S. (1997). *The Endangered Atmosphere: Preserving a Global Commons. Columbia*, NY: University of South Columbia Press.

Stevis, D. and Assetto, V. J. (eds) (2001). T*he International Political Economy of the Environment*. Boulder, CO: Lynne Rienner.

Steffen, W., Sanderson, A., Tyson, P. D., Jäger, J., Matson, P. A., Moore III, B., Oldfield, F., Richardson, K., Schellnhuber, H. J., Turner, B. L. and Wasson, R. J. (2004). *Global Change and the Earth System: A Planet under Pressure*. New York: Springer.

Steffen, W., Grinewald, J., Crutzen, P. and McNeill, J. (2011). The Anthropocene: From Global Change to Planetary Stewardship. *Ambio*, 40, 739-761.

Steffen, W., Richardson, K., Rockström, J., Cornell, S. E., Fetzer, I., Bennett, E. M., Biggs, R., Carpenter, S. R., de Vries, W., de Wit, C. A., Folke, C., Gerten, D., Heinke, J., Mace, G. M., Persson, L. M., Ramanathan, V., Reyers, B. and Sörlin, S. (2015). Planetary Boundaries: Guiding Human Development on a Changing Planet. *Science*, 347(6223), 736-746.

Suliman, M. (ed.) (1999). *Ecology, Politics and Violent Conflict*. London: Zed.

Thomas, W. (ed.) (1956). *Man's Role in Changing the Face of the Earth*. Chicago, IL: University of Chicago Press.

Turco, R., Toon, O. B., Ackerman, T. P., Pollack, J. B. and Sagan, C. (1983). Nuclear Winter: Global Consequences of Multiple Nuclear Explosions. *Science*, 222(4630), 1283-1292.

United Nations Environment Programme (2012). *Global Environmental Outlook GEO 5: Environment for Development*. Nairobi: UNEP.

Vogler, J. (1995). *The Global Commons: A Regime Analysis*. Chichester: Wiley.

Wapner, P. (1996). *Environmental Activism and World Civil Politics*. New York: State University of New York.

Ward, B. and Dubos, R. (1972). *Only One Earth*. Harmondsworth: Penguin.

Webersik, C. (2010). *Climate Change and Security: A Gathering Storm of Global Challenges*. Santa Barbara, CA: Praeger.

Welzer, H. (2012). *Climate Wars*. Cambridge: Polity.

World Commission on Environment and Development (1987). *Our Common Future*. Oxford: Oxford University Press.

Young, O. (1994). *International Governance: Protecting the Environment in a Stateless Society*. Ithaca, NY: Cornell University Press.

4

The Institutionalisation of Climate Change in Global Politics

NINA HALL
HERTIE SCHOOL OF GOVERNANCE, GERMANY

*'Humanity is conducting an unintended, uncontrolled, globally
pervasive experiment whose ultimate consequences
could be second only to nuclear war.'*
1988 World Conference on the Changing Atmosphere

Today it is commonplace to state that climate change is an urgent global priority. States, and scientists, have highlighted its destructive effects. In fact, scientific studies abound illustrating how climate change will lead to an increased frequency of extreme weather events, triggering more intense storms, melting polar icecaps and glaciers and raising sea levels (IPCC, 2014). It will have major effects on everything from agriculture to the spread of diseases. Yet anthropogenic climate change was once dismissed by many scientists, ignored by heads of state and seen as irrelevant by our multilateral institutions. So how has climate change become a top global priority? And how do we know that it will continue to be so?

This contribution argues that climate change has become institutionalised in global affairs as a top priority issue. First, there is a strong scientific consensus that greenhouse gas emissions are increasing due to human behaviour and this is driving up average global temperatures. In addition, states, including major powers, regularly meet and discuss how to mitigate climate change at global summits. Third, states have committed significant new resources to address climate adaptation and mitigation in developing countries. Fourth, a wide range of multilateral institutions from the United Nations High Commissioner for Refugees (UNHCR) to the World Health Organisation (WHO) have institutionalised climate change within their work. In addition, a transnational civil society movement for climate justice has also been critical at keeping pressure on states and global institutions to take action, although this is not the focus here (Hadden, 2015). This article complements our understanding of how environmental issues become institutionalised in global affairs (see Falkner, 2012).

This chapter argues that climate change is now widely recognised by states and institutions as one of the top global challenges. Change has occurred along four dimensions: 1) scientific consensus; 2) political action; 3) financial resources; and 4) institutionalisation of climate change in multilateral organisations. The chapter draws on an examination of G7 and G8 communiqués as well as extensive research on international organisations engagement with the United Nations Framework Convention on Climate Change (UNFCCC) and climate change (Hall, 2015).

Scientific consensus on climate change

In the 19th and 20th centuries, a series of scientific studies made the case that humans, through industrialisation, were affecting climate change. Already in 1859 John Tyndall proved the 'greenhouse effect' by demonstrating that gases have different absorption patterns (Paterson, 1996). In 1938, Guy Stewart Callendar found that increasing concentration of carbon dioxide in the atmosphere was linked to an increase in world temperature (Hulme, 2009). Initially other scientists did not take these results seriously, doubting that carbon dioxide levels had increased. Furthermore, Callendar presented his findings just as world attention was on the rising power of Nazis in Germany and the lead-up to World War II.

In the second half of the 20th century scientific evidence for climate change grew. In the 1950s and 1960s, scientists began modelling carbon dioxide levels and found further evidence of anthropogenic climate change (Paterson, 1996: 22; Hulme, 2009). In 1979, scientists met at the World Climate Conference, one of the first international conferences dedicated to climate change. Legislators also started to listen to scientific concerns: in 1988, James Hansen, a scientist for the National Atmospheric and Space Administration (NASA), gave evidence in a United States Senate hearing on the dangers and likelihood of global warming. In 1988, the first intergovernmental conference on climate change was held in Toronto and attended by many scientists and politicians. The conference recommended a 20 per cent reduction in carbon dioxide emissions by 2005 and the establishment of an inter-governmental scientific body to monitor the issue: the Inter-Governmental Panel on Climate Change (IPCC).

The IPCC was tasked with preparing a 'comprehensive review and recommendations with respect to the state of knowledge of the science of climate change; social and economic impact of climate change, possible response strategies and elements for inclusion in a possible future international convention on climate' (IPCC, 2015). In 1990, the IPCC published its first report outlining the scientific evidence for anthropogenic climate change (IPCC, 1990). Since then, they have issued dozens of additional reports, the work of thousands of scientists who peer review each other's work, and have become the global authority on climate change. The Fifth IPCC Report, released in 2014, emphasised the strong scientific case for anthropogenic climate change, stating that the 'warming of the climate system is unequivocal' (IPCC, 2014). The IPCC previously co-won the Nobel Peace Prize in 2007 with Al Gore for their 'efforts to build up and disseminate greater knowledge about man-made climate change and to lay the foundations for the measures that are needed to counteract such change' (IPCC, 2015).

Scientific knowledge by its nature is always open for debate and contestation. For example: the IPCC has not always been correct in its predictions. In a 2007 report they claimed incorrectly that Himalayan glaciers would melt away by 2035 (IPCC, 2010). Modelling the impacts of climate change is challenging, hence it is difficult to predict the exact impacts in a given locale. However, there is now a clear consensus that greenhouse gas emissions (caused by the burning of fossil fuels which is the basis of industrialised economies) has led to an increase in the global average temperature. The increase in average global temperature is having a number of other effects: from the melting of the polar icecaps and glaciers to an increased frequency and intensity of storms and drought in many areas of the world. Furthermore, there are likely to be critical tipping points which can lead to irreversible changes (Lenton, 2011). Over the past 150 years climate change has gone from an issue dismissed by many scientists to being widely accepted as a critical global challenge which national leaders must respond to.

Political action on climate change

Since the late 1980s world leaders have acknowledged the potential disastrous impacts of climate change. In 1988, British prime minister Margaret Thatcher made a speech to the Royal Society of London in which she drew attention to climate change, claiming that it is possible 'we have unwittingly begun a massive experiment with the system of this planet itself'

(Hulme, 2009: 65). In the same year, the foreign minister of the Soviet Union, Eduard Shevardnadze, also called for action on climate change in a speech to the UN General Assembly, and, during his election campaign, President George H. W. Bush pledged to hold a global conference on climate change at the White House (Paterson, 1996: 35). In 1989, the Group of Seven (G7), the Non-Aligned Countries meeting and the Commonwealth Heads of Government meeting all stated that global warming was a pressing global issue.

However, leaders in the late 1980s and early 1990s predominantly saw climate change as one of a long list of environmental issues, not as the single most important global environmental issue, as it has now become. Leaders – even ones not known for their progressive politics such as Thatcher – who highlighted the impacts of climate change did so in the context of an increased global awareness of environmental problems. In 1992, states met in Rio de Janeiro at the UN Conference on the Environment and Development (the 'Earth Summit'). It was the largest global environmental meeting since the Stockholm Environmental Conference in 1972 – when states acknowledged their duty to protect and improve the environment at an international summit for the first time (Falkner, 2012: 513).

In the lead-up to and during the 1992 Rio Earth Summit conference, world leaders highlighted a number of environmental problems including biodiversity, the growing ozone hole, pollution, desertification and climate change. G7 and G8 communiqués reflect the perception of climate change as one of many important global environmental problems. In 1987, for instance, the G7 communiqué argued for 'further action' on 'global climate change, air, sea and fresh water pollution, acid rain, hazardous substances, deforestation, and endangered species'. Climate change was not considered a stand-alone priority issue, but a subset of other major global environmental problems.

This began to change with the establishment of the UNFCCC, which was opened for signature in 1992. The UNFCCC aimed to stabilise greenhouse gas 'concentrations in the atmosphere at a level that would prevent dangerous interference with the climate system' (United Nations, 1992). The initial goal was to stabilise emissions at 1990 levels by 2000 and the UNFCCC became the forum where states negotiated how to reach this target. The first annual negotiations – or Conference of the Parties (COP) – were held in Berlin in 1995 and state parties agreed that industrialised states would need to make binding commitments to reduce emissions. The UNFCCC institutionalised climate change and ensured that states would regularly meet to discuss how to address growing global greenhouse gas emissions.

At the UNFCCC meetings states staked out their positions on climate change – some such as Saudi Arabia were sceptics (Depledge, 2008) and others, in particular the small island developing states, demanded urgent action. There is a growing body of International Relations scholarship that examines the evolution of states' positions (Torney, 2015); the formation of coalitions predominantly along North–South lines; and negotiations over various agreements (Barnett, 2008; Roberts, 2011). By 1997, more than 150 countries agreed to sign the Kyoto Protocol which binds most industrialised states and economies in transition to reduce greenhouse gas emissions (UNFCCC, 2015b). They are known as 'Annex I' countries. The Protocol took a further four years to be operationalised as UNFCCC negotiations collapsed in 2000 over major disagreements between the US and the European Union (EU). Then in 2001, the new US president, George W. Bush, announced he would withdraw the United States from the Kyoto Protocol, which President Clinton had previously signed (Busby, 2010). The absence of the world's largest economy and emitter jeopardised an agreement; however, other states continued to negotiate and, in 2001, finalised the Kyoto Protocol.

In the 1990s and early 2000s the UNFCCC was the main forum for states to discuss climate change. This changed in the mid-2000s, as world powers made climate change a stand-alone agenda item in the important global economic and security summits. By 2005 climate change was one of the top agenda items at the G8 summit agenda in Gleneagles. The United Kingdom, host of the summit, also invited five 'emerging' states (Brazil, China, India, Mexico and South Africa) to attend. They formed a new group G8+5 to build an agreement on climate change and issued a separate statement, the Gleneagles Plan of Action 'setting out our common purpose in tackling climate change' (G8 Chair, 2005). The communiqué stated that 'all of us agreed that climate change is happening now [...] and resolved to take urgent action to meet the challenges we face' (G8 Chair, 2005). Subsequent meetings of the G7/G8 continued this focus on climate change, which was seen as an important issue that warranted discussion beyond the UNFCCC and by heads of the world's most powerful economies.

In addition, states began to see climate change as not only an environmental issue, but also an economic issue. This shift in perception of climate change was facilitated by the United Kingdom's Stern Review. Gordon Brown, UK chancellor in 2006, commissioned Lord Nicholas Stern, a prominent economist, to write a report on the costs of climate change. The report made a strong case for immediate emissions reductions on the basis that the short-term costs of mitigation would be significantly less than the long-term costs of inaction (Stern, 2006). The report had a major international impact as it was the first report commissioned by a government to make an economic case for emissions reductions and Lord Stern, backed by the UK government, disseminated this message widely in late 2006 and 2007 (Torney, 2015).

Climate change was also seen as a threat to security. Some states and many civil society organisations, non-governmental organisations and academics argued that climate change would lead to an increase in conflict, be a new driver of displacement and make some small island states uninhabitable (Myers, 1993; 1997). In fact, the UK successfully campaigned for the UN Security Council to debate climate change in April 2007. A record number of states spoke during this meeting – 55 states; 40 non-members – and outlined the urgency of addressing climate change because of its potential threats to security (United Nations, 2007). In short, by 2007, both the world's premier economic and security forum had made climate change an explicit top priority, singled out and above other environmental issues.

The 2009 UNFCCC summit in Copenhagen was one of the largest gatherings of world leaders ever. All the world's eyes turned to Denmark to see if states could come up with a new fair and binding treaty to mitigate carbon emissions. It was a remarkable moment for global politics: almost every head of state spoke at the negotiations in the Bella Centre. In the final hours, US president Barack Obama drew up an agreement with the leaders of China, India, Brazil and South Africa; but not all states agreed to their plan after hours of negotiating through the night. The conference finally emerged with an agreement that all member states were invited to 'take note of' but was not officially endorsed by all UNFCCC states (UNFCCC, 2009). Copenhagen was widely perceived as a failure. However, negotiations did make more progress the following years at Cancun, Durban and Warsaw. For example, a new global climate fund (GCF) was established to finance mitigation and adaptation in developing countries.

Immediately after Copenhagen, interest in climate change ebbed, in part due to disillusionment with the UNFCCC process. World leaders also shifted their attention to the 2012 Rio+20 World Environmental Conference. However, in the past two years world powers have again prioritised climate change at major global summits and made significant commitments to reduce their carbon emissions. In November 2014, for instance, US president Barack Obama met with President Xi Jinping of China and both made new commitments to reduce their national carbon emissions, paving the way for other states to follow suit. Obama announced a new target to cut net greenhouse gas emissions by 26–28 per cent below 2005 levels by 2025 and Xi announced targets to peak carbon dioxide emissions around 2030 with the intention of peaking earlier, and increasing non-fossil fuel share of all energy to around 20 per cent by 2030 (The White House – Office of the Press Secretary, 2014). In September 2015 they both reaffirmed their commitments to reach an ambitious agreement at the UNFCCC summit in Paris. The fact that the US and China – the two major world powers of the 21st century – made climate change a central part of their bilateral negotiations signals the importance of the issue internationally today.

Meanwhile, in July 2015 Germany made climate change a core focus of the G7 Summit at *Schloss Elmau* and the final summit communiqué emphasised that,

> deep cuts in global greenhouse gas emissions are required with a decarbonisation of the global economy over the course of this century [...] We commit to doing our part to achieve a low-carbon global economy in the long-term including developing and deploying innovative technologies striving for a transformation of the energy sectors by 2050 and invite all countries to join us in this endeavour (G7, 2015: 15).

Heads of states from major world powers to those most affected by climate change have prioritised the issue, made significant shifts in their positions and committed to taking action on climate change. We saw the most compelling example of this in Paris in December 2015 when states forged a new international agreement on climate change. In the Paris Agreement states agreed to keep average global temperature increases below 2 degrees, with the aim of keeping increases within 1.5 degrees. They also laid out a clear process to reach this goal: every five years they will submit more ambitious plans laying out how they will reduce their greenhouse emissions. However, it is worth noting that states intended nationally determined contributions (INDCs) do not meet the two-degree global warming target (for a full list of INDCs, see UNFCCC, 2015a). We still need to see further cuts to stop dangerous climate change.

Financing for climate change

In the 2000s, states also institutionalised climate change as a top priority in global affairs by committing significant new resources to it. The first climate financing was established in Rio in 1992 (Mingst and Karns, 2007: 216). The Global Environment Facility (GEF) channelled grants from developed to developing states to address biodiversity, climate change, ozone layer depletion and international waters (Young, 2002). The GEF enabled the United Nations Development Programme (UNDP), the United Nations Environment Programme (UNEP) and the World Bank (the only three multilaterals who could access it) to expand their environmental and climate change activities (Hall, forthcoming/a; forthcoming/b).

Subsequently, since the turn of the millennium, state parties to the UNFCCC established a series of new and explicitly climate change orientated financing mechanisms. In 2000 at the 6th annual UNFCCC summit, as the negotiations over Kyoto became difficult, the EU agreed to establish an annual climate change fund of US$15 million to target adaptation as well as mitigation. Subsequently at the next COP in Marrakech in 2001, three multilateral funds were established: the Special Climate Change Fund (SCCF), based on voluntary donations to facilitate technology transfer from developed to developing states; the Least Developed Countries Fund (LDCF) for least developed countries to develop National Adaptation Programmes of Action (NAPA); and the Adaptation Fund, which was financed by a 2 per cent levy on the Clean Development Mechanism (CDM). The establishment of these three climate funds offered new financing opportunities for multilateral organisations. They were also important as they shifted climate change activities from purely focusing on reducing carbon emissions (mitigation) to acknowledging that developing states would need assistance to prepare for and deal with the impacts of climate change (adaptation).

A major windfall of new financing was announced in 2009 at the UNFCCC summit in Copenhagen. Donor states committed to significant 'new and additional' climate financing (UNFCCC, 2009). This financing would come in two forms: first, a new 'fast-track fund' for the 2010–2012 period, totalling up to US $30 billion per annum. Second, states committed to mobilising new financing of up to US$100 billion by 2020 from a range of private and public sources. Some of this financing would flow through the new Green Climate Fund. States have begun to commit significant resources to the GCF. In September 2014, 125 heads of state and government as well as 800 leaders from business, finance and civil society attended a UN Climate Summit and pledged support totalling up to US$2.3 billion for the Green Climate Fund. Subsequently, in mid-2015 Germany announced it would double its climate finance to €4 billion a year by 2020, China declared it would provide US$3.1 billion in climate finance, the United Kingdom announced it will provide £5.8 billion between 2016 and 2021, and France €5 billion a year by 2020 (World Resources Institute, 2015). If all these pledges are fully paid, the GCF will be the largest multilateral climate fund (Heinrich Boell Foundation, 2015). However, as of October 2015, the fund was still not fully operational.

The growth of climate finance is an important trend in international relations. It means developing countries have resources to adapt to and mitigate climate change. However, climate finance is not clearly 'new and additional' from overseas development assistance, as originally pledged at Copenhagen (Stadelman et al., 2010). In fact, many donor states are refocusing their existing development budgets to prioritise climate mitigation and adaptation. The growth of climate finance has also enabled multilateral banks, and many international development organisations, to expand their work on climate mitigation and adaptation. Many international organisations, with no established mandate for climate adaptation or mitigation, have established new departments, teams and projects to target climate change as will be discussed next (Hall, forthcoming/b).

Multilateral institutionalisation of climate change

International development and humanitarian organisations are at the forefront of climate change. They assist the most vulnerable countries to deal with and prepare for droughts, famines and other natural disasters. Yet most of our existing international organisations were established in the first half of the 20th century – when climate change was neither a global priority nor a scientific reality. The World Health Organisation (WHO), UNICEF, International Organisation for Migration (IOM), UNHCR and other international organisations thus had no original mandate to respond to climate change. Over the past two decades there has been a remarkable shift as many multilateral institutions have engaged in the UNFCCC negotiations, accessed climate funds and developed new programmes and policies on adaptation and mitigation.

First, many more international organisations are engaging with the UNFCCC. The number of international organisations attending the annual climate negotiations has more than doubled between 1994 and 2009 (see Hall, 2015). Peak attendance was at the Copenhagen negotiations in 2009, when over 100 intergovernmental organisations attended, compared with 42 in 1994 at COP1. The range of international organisations has also expanded beyond development and environment organisations, to humanitarian, refugee, migration, and health organisations (Hall, forthcoming/a).

Take the UNHCR as an example. This organisation was established in 1951 to assist refugees, defined as someone with 'a well-founded fear of persecution based for reasons of race, religion, nationality, membership of a particular social group or political opinion, is outside his country of nationality and is unable or owing to such fear, is unwilling to avail himself of the protection of that country' (UNHCR, 1951). It had no mandate to help those displaced by natural disasters such as floods or droughts (Betts et al., 2012). Yet there have been calls for this organisation to expand its mandate and encompass people affected by natural disasters and forced to flee across borders due to climate change (Biermann and Boas, 2010). Although it does not have a mandate to respond to the latter, it has broadened its focus in the past decade. UNHCR often assists internally displaced persons (IDPs) after natural disasters; such was the case in Pakistan after the 2010 floods and in

2009 after Cyclone Nargis in Myanmar. International organisations are adapting their tasks and mandates to meet new demands.

Other humanitarian organisations have also become more engaged with climate change as the UNFCCC negotiations broadened their focus from mitigation to adaptation (Hall, 2015; forthcoming/b). In the 1990s and early 2000s, when climate change was primarily about how to reduce emissions; humanitarian organisations such as UNHCR, IOM and the International Committee of the Red Cross did not engage with climate change. However, when it was acknowledged that climate change was already having a major impact on the most vulnerable countries and likely to lead to more humanitarian (natural) disasters, the humanitarian community became involved. Humanitarian organisations established a special task force under the Inter-Agency Standing Committee to explore how to address climate change in humanitarian situations and wrote a number of submissions to the UNFCCC (Hall, forthcoming/a).

In another telling example, Margaret Chan, director general of the World Health Organisation (WHO), now identifies the climate deal in Paris as the 'most important health agreement of the century' (Climate Change Policy and Practice, 2015). This is because there is 'overwhelming evidence' that climate change endangers human health and we need 'decisive action' to change the trajectory of increased emissions and thus reduce costs on the health system and community. The Sustainable Development Goals (SDG), announced in September 2015, have also entrenched climate change as a core priority for all development organisations. Goal 13 is to 'take urgent action to combat climate change and its impacts' (UN General Assembly, 2015).

Crucially, in the last decade there is an awareness of how climate change spills over into other many other issue areas. It can no longer be dealt with in the UNFCCC alone, and we are seeing the emergence of a 'regime complex' (Keohane and Victor, 2011), in which many global institutions are involved. These institutions will continue to act on climate change because of humanitarian and development needs. In addition, there is vast financing being set aside and many multilateral institutions have established new teams, programmes and some have reprioritised climate change as a central focus within their mandate (such as UNDP) (Hall, 2015: 84).

Conclusion

Climate change is a major political, economic, and social issue that has become institutionalised in global affairs. This has happened because of an increased scientific and political consensus. We now see climate change being discussed at major forums from the G7 to the UN Security Council on a regular basis. This was not the case twenty years ago. Major powers have made it a priority in their bilateral discussions – such as the November 2014 summit between the presidents of China and the US. They have also committed significant financing to address mitigation and adaption. There is a growing awareness that climate change is impacting many states now, particularly the most vulnerable developing countries and low-lying island states. Multilateral institutions from the UNHCR to WHO are also prioritising it within their mandates and assisting developing states cope with its effects. Climate change will not go away from international relations because it is institutionalised at this level.

So why we have not yet resolved climate change, given the high political attention and resourcing it has received in recent decades? Unfortunately, reducing greenhouse gas emission requires great political will and profound transformations in our global economy and we are just at the beginning of this process. We need continued action on all four fronts – financing, multilateral organisations, heads of state and scientific research – as well as concerted action from civil society to decouple economic growth from greenhouse gas emissions. The Paris agreement was a positive step-forward in this direction.

The author would like to thank Diarmuid Torney, Olivia Gippner, Steffen Lohrey and Lisa Schmid for their invaluable feedback on this chapter.

References

Barnett, J. (2008). The Worst of Friends: OPEC and G77 in the Climate Regime. *Global Environmental Politics*, 8(4), 1-8.

Betts, A., Loescher, G. and Milner, J. (2012). *UNHCR: The Politics and Practice of Refugee Protection*. Abingdon: Routledge.

Biermann, F. and Boas, I. (2010). Preparing for a Warmer World: Towards a Global Governance System to Protect Climate Refugees. *Global Environmental Politics*, 10(1), 60-88.

Busby, J. (2010). *Moral Movements and Foreign Policy*. Cambridge: Cambridge University Press.

Climate Change Policy and Practice. (2015). *WHO Calls for Action to Protect Health from Climate Change*. Retrieved from http://climate-l.iisd.org/news/who-calls-for-action-to-protect-health-from-climate-change/

Depledge, J. (2008). Striving for No: Saudi Arabia in the Climate Change Regime. *Global Environmental Politics*, 8(4), 9-35.

Falkner, R. (2012). Global Environmentalism and the Greening of International Society. *International Affairs*, 88(3), 503-522.

G7 (1987). *Leaders Summit Communiqué – Venice*. Retrieved from http://www.library.utoronto.ca/g7/summit/1987venice/communique/index.html

G7 (2015). *Leaders' Declaration: G7 Summit*. Retrieved from https://www.whitehouse.gov/the-press-office/2015/06/08/g-7-leaders-declaration

G8 Chair. (2005). *Chair's Summary – Gleneagles*. Retrieved from http://www.g8.utoronto.ca/summit/2005gleneagles/summary.html

G8 Chair (2007). *Chair's Summary – Heiligendamm*. Retrieved from http://www.g8.utoronto.ca/summit/2007heiligendamm/index.html

Hadden, J. (2015). *Networks in Contention, the Divisive Politics of Climate Change*. New York: Cambridge University Press.

Hall, N. (2015). Money or the Mandate? Why International Organizations Are Engaging with the Climate Change Regime. *Global Environmental Politics*, 15, 79-96.

Hall, N. (forthcoming/a). A Catalyst for Cooperation? The Inter-Agency Standing Committee and the Humanitarian Response to Climate Change. *Global Governance*.

Hall, N. (forthcoming/b). *Displacement, Development and Climate Change: International Organizations Moving beyond their Mandates*. Oxford and New York: Routledge.

Heinrich Boell Foundation. (2015). *Making the Green Climate Fund 'Effective' Soon – in a Lasting Way*. Retrieved from https://us.boell.org/2015/05/05/making-green-climate-fund-effective-soon-lasting-way

Hulme, M. (2009). *Why We Disagree About Climate Change, Understanding Controversy, Inaction and Opportunity*. Cambridge, MA: Cambridge University Press.

IPCC (1990). *Climate Change, the IPCC Scientific Assessment*. Cambridge: Cambridge University Press. Retrieved from https://www.ipcc.ch/publications_ and_data/publications_ipcc_first_assessment_1990_wg1.shtml

IPCC (2010). IPCC *Statement on the Melting of Himalayan Glaciers*. Retrieved from http://www.ipcc.ch/pdf/presentations/himalaya-statement-20january2010.pdf

IPCC (2014). Climate Change 2014 – *5th Assessment Report*. Retrieved from http://www.ipcc.ch/pdf/assessment-report/ar5/syr/SYR_AR5_FINAL_full.pdf

IPCC (2015). *Organization History*. Retrieved from http://www.ipcc.ch/ organization/organization_history.shtml

Keohane, R, O. and Victor, D. G. (2011). The Regime Complex for Climate Change. *Perspectives on Politics*, 9(1), 7-23.

Lenton, T. M. (2011). Early Warning of Climate Tipping Points. *Nature Climate Change*, 1(4), 201-209.

Mingst, K. A. and Karns, M. P. (2007). *The United Nations in the 21st Century*. Cambridge: Westview Press.

Myers, N. (1993). Environmental Refugees in a Globally Warmed World. *BioScience*, 43(11), 752-761.

Myers, N. (1997). Environmental Refugees. *Population and Environment: A journal of Interdisciplinary Studies*, 19, 167-182.

Paterson, M. (1996). *Global Warming and Global Politics*. London: Routledge.

Roberts, J. T. (2011). Multipolarity and the New World (Dis)Order: US Hegemonic Decline and the Fragmentation of the Global Climate Regime. *Global Environmental Change*, 21, 776-784.

Stadelman, M. J., Roberts, J. T. and Huq, S. (2010). Baseline for Trust: Defining 'New and Additional' Climate Funding. London: International Institute for the Environment and Development. Retrieved from http://pubs.iied.org/pdfs/17080IIED.pdf

Stern, N. (2006). *The Stern Review on the Economics of Climate Change*. Retrieved from http://webarchive.nationalarchives.gov.uk/+/http:/www.hm-treasury.gov.uk/independent_reviews/stern_review_economics_climate_change/stern_review_report.cfm

The White House, Office of the Press Secretary (2014). *Fact Sheet: US-China Joint Announcement on Climate Change and Clean Energy Cooperation*. Retrieved from https://www.whitehouse.gov/the-press-office/2014/11/11/fact-sheet-us-china-joint-announcement-climate-change-and-clean-energy-c

Torney, D. (2015). *European Climate Leadership in Question: Policies toward China and India*. Cambridge, MA: MIT Press.

United Nations (1992). *UN Framework Convention on Climate Change*. Retrieved from http://unfccc.int/files/essential_background/background_publications_htmlpdf/application/pdf/conveng.pdf

United Nations (2007). *Chair's Final Comments, UN Security Council 5663rd Meeting, S/Pv. 5663*.

UN General Assembly. (2015). Transforming *Our World: The 2013 Agenda for Sustainable Development*. Retrieved from http://www.un.org/ga/search/view_doc.asp?symbol=A/70/L.1&Lang=E

UNFCCC (2009). *Copenhagen Accord*. Retrieved from http://unfccc.int/resource/docs/2009/cop15/eng/11a01.pdf

UNFCCC (2015a). *Intended Nationally Determined Contributions*. Retrieved from http://www4.unfccc.int/submissions/indc/Submission%20Pages/submissions.aspx

UNFCCC (2015b). *Parties and Observers*. Retrieved from http://unfccc.int/parties_and_observers/items/2704.php

UNHCR (1951). *Convention and Protocol Relating to the Status of Refugees*. Retrieved from http://www.unhcr.org/3b66c2aa10.html

World Resources Institute (2015). *Statement: WRI's Andrew Steer Welcomes France's Climate Finance Commitment*. Retrieved from http://www.wri.org/news/2015/09/statement-wri%E2%80%99s-andrew-steer-welcomes-france%E2%80%99s-climate-finance-commitment

Young, Z. (2002). *A New Green Order? The World Bank and the Politics of Global Environment Facility*. London: Pluto Press.

5

Refusing to Acknowledge the Problem: Interests of the Few, Implications for the Many

KIRSTI M. JYLHÄ
UPPSALA UNIVERSITY, SWEDEN

Human-induced climate change is a major threat for people and other inhabitants of Earth (Intergovernmental Panel on Climate Change [IPCC], 2014). Climate scientists are highlighting the importance of mitigation efforts that are needed to avoid the most severe consequences, but many people do not have interest in the issue and some even deny that climate is changing due to human activities (Leiserowitz et al., 2013). Given the widespread scientific evidence, it is important to move on from questioning whether the climate is changing due to human actions to asking what hinders people from acknowledging it.

Individuals may deny climate change for various reasons (see American Psychological Association, 2009; Milfont, 2010; Ojala, 2012). For instance, some find it hard to comprehend the problem due to its complexity and some deny it as an effort to cope with negative feelings that fear of climate change evokes in them. Also, scientific conclusions are not reported as definite truths, but rather in terms of likelihood and probabilities. It may be difficult for lay people to interpret conclusions that are reported in this way, which can lead them to underestimate the level of certainty in the predictions and consensus among climate scientists (Budescu et al., 2009).

Climate scepticism is also tactically promoted by organised 'denial machines' that are funded by wealthy foundations and corporations (Oreskes and Conway, 2010; McCright and Dunlap, 2011). These denial machines aim to influence public opinion by manufacturing uncertainty and doubt. Two of their main strategies consist in attacking climate science and scientists, and spreading counterevidence about climate change. In part because of this influence, strongly dismissive views on climate change are repeatedly being presented in the media and everyday discussions, hindering public support and delaying environmental action. Of particular interest for this paper, the majority of the literature providing counterevidence for climate change is published outside scientific communities and has links to politically conservative movements and think tanks (Jacques et al., 2008). This counterevidence also gains more support from conservative voters, and is more commonly communicated through conservative media and blogs, when compared to liberals (McCright and Dunlap, 2011; Feldman et al., 2014). Thus, although the reasons for denial may vary, political orientation seems to be a central issue.

Political ideology and climate change denial

Substantial evidence from different countries shows that politically conservative/right-wing individuals report higher levels of climate change denial when compared to their liberal/left-wing counterparts (McCright and Dunlap, 2011; Poortinga et al., 2011; Häkkinen and Akrami, 2014; McCright et al., 2016; Milfont et al., 2015). This divide has been reported not only when it comes to denying the observed and predicted changes in the climate system but also when it comes to denying human contribution to these changes, as well as the danger and seriousness of them.

One reason for the ideological divide is that conservative voters are exposed to more dismissive messages about climate change, as conservative politicians and other role models have been communicating more sceptical views on climate change than their liberal counterparts (Jacques et al., 2008). However, it is important to note that opinions are not only dependent on external influences such as exposure to different kinds of ideological messages. Rather, there are also certain psychological factors (as are discussed in this contribution) that make individuals more or less prone to adopt different ideological views (Jacquet et al., 2015). Thus, conservatives can be expected to be inclined to doubt the reality of climate change even if they have not been influenced by any ideological messages regarding the issue. Indeed, the observation that climate change denial is largely promoted by conservative think tanks supports the suggestion (Jacques et al., 2008).

Climate change is an increasingly political issue in part due to ideological communications, but the psychological factors that underpin political orientation could explain what led some conservatives to campaign against climate science in the first place. In this chapter, I discuss climate change denial in relation to two overall psychological tendencies that are linked to political orientation – resistance to societal change and acceptance of inequality – as well as psychological mechanisms that underpin these tendencies.

Resistance to change

Preference for traditional lifestyle and values, as well as resistance to social and economic change, is a core component of conservative ideologies (Jost et al., 2003). When compared to liberals, conservatives also tend to favour system-justifying ideologies (see Jost et al., 2003), defined as acceptance and defence of the status quo, such as the prevailing social and economic structures and norms (Jost and Banaji, 1994). Importantly, individuals who are motivated to perceive the status quo as legitimate and desirable resist information about environmental problems caused by our current lifestyle (Feygina et al., 2010).

One reason for the attractiveness of conservative and system-justifying ideologies is that they enable relatively simple and certain ways of explaining various phenomena and offer clear and stable guidelines for handling different situations (Jost et al., 2003). Thereby, they provide a sense of certainty, stability and safety and reduce anxious and negative feelings. Put another way, conservative ideology can be considered as a motivated cognition that satisfies the need to manage uncertainty and threat (Jost et al., 2003). This view is supported by consistent findings showing that a tendency to see the world as a dangerous place and motivation to avoid uncertainty and threat is more common among conservatives than liberals (Jost et al., 2003; 2007).

Motivation to avoid uncertainty is of importance when explaining climate change denial. More specifically, climate change is a complex phenomenon that cannot be explained and predicted with full certainty. In order to cope with this uncertainty, individuals may be attracted to the simplest and most definite explanation that has been given for climate change; that is, 'climate change is not occurring at all'. As conservatives tend to dislike uncertainty more than liberals do (Jost et al., 2003), this uncertainty avoidance tendency can be expected to be more common among conservatives than liberals. As for threat avoidance, climate change might be perceived as a twofold threat for people: it is a threat to life on Earth, while climate change mitigation is a threat to the status quo. Denial offers a way to cope with both these threats, as it diminishes the fear for climate change and enables people to perceive the status quo as unchangeable and justifiable again (Feygina et al., 2010;

Ojala, 2012). However, no study has investigated whether the motivation to manage uncertainty and threat indeed explains any part of the relation between political ideology and climate change denial.

Acceptance of inequality and environmental injustice

As discussed above, one reason for climate change denial is that people are motivated to accept the status quo and adhere to traditional ways of living. However, is it not a huge risk to ignore the warnings about dangerous climate change simply out of motivation to continue living as before? The answer to this question seems to be 'no' for some people, who do not consider themselves or their loved ones to be at risk (Milfont, 2010; Spence et al., 2012). Rather, they are inclined to distance themselves from the problem and believe that climate change affects people who are psychologically and geographically distant from them and that its consequences will be felt more in the future than today.

These perceptions are somewhat accurate for some, as climate change is not likely to affect most seriously the wealthy and powerful populations (IPCC, 2014). Rather, disadvantaged people and nations are facing the highest and most acute risks, as they lack the needed resources to cope with the negative effects, such as reduced food and water supplies and extreme weather events. Also, future generations and non-human animals are at serious risk (IPCC, 2014). The populations least responsible for the current greenhouse gas emission will be facing the most serious consequences of climate change. What is more, climate change is likely to slow down economic growth, exacerbate poverty and create new poverty traps (IPCC, 2014). Thus, climate change can be perceived as a form of social injustice (Schlosberg, 2013), which offers an important further explanation for why political ideology is linked to climate change denial.

When compared to liberals, politically conservative and system-justifying individuals tend to accept policies that maintain inequality and injustice (Jost et al., 2003) and also score higher in a variable called social dominance orientation (SDO) (Jost and Thompson, 2000; Wilson and Sibley, 2013) that captures acceptance and promotion of group-based social hierarchies and dominance (Pratto et al., 1994). Recent research suggests that individuals who report high levels of SDO support human dominance over the rest of nature, accept nature utilisation and environmentally harmful actions (particularly if such actions benefit high-status groups) and deny climate change (Jackson et al., 2013; Milfont et al., 2013; Dhont et al., 2014; Milfont and Sibley, 2014; Jylhä and Akrami, 2015). Thus, SDO could help to explain the relation between political orientation and climate change denial.

In order to test this, Jylhä and Akrami (2015) have investigated whether the relation between conservative ideology and climate change denial holds after the effect of this social dominance orientation is statistically taken into account. It was found that the effects of conservative ideologies (i.e. political orientation, system justification and right-wing authoritarianism) on denial either vanish or substantially decrease when SDO is controlled (Häkkinen and Akrami, 2014). In other words, SDO explains why some individuals are denying climate change, and above the effect of SDO, the other ideological variables add only a small or zero contribution to explaining denial. Thus, an important explanation for the relation between political ideology and climate change denial is that conservatives accept and promote inequalities to a higher degree when compared to liberals (see also McCright and Dunlap, 2011). This suggests that it could be beneficial to focus specifically on SDO when explaining the ideological bases behind climate change denial, rather than focusing on political ideology or conservatism in general.

Social dominance orientation and climate change denial

SDO measures positive views on social hierarchies, and recent research has demonstrated that this tendency extends into accepting hierarchical relations between humans and nature as well (Milfont et al., 2013). In this hierarchical system, humans are perceived as a superior group with a legitimate right to dominate the rest of the nature. In line with these findings, Jylhä and Akrami (2015) have found that SDO correlates with accepting attitudes regarding nature dominance. They have also shown that acceptance of these two types of group-based dominance – social and nature dominance – uniquely predict climate change denial. Although future studies should investigate this question further, it seems that climate change is disputed as an effort to defend the existing social and human–nature hierarchies.

Psychological factors that are linked to SDO offer further understanding about climate change denial. Individuals who score high in SDO tend to perceive the world as a 'competitive jungle' where hierarchies are inevitable and natural (Duckitt, 2001). Also, people may learn to hold desirable views on power structures through socialisation processes, because such views are widespread in society (Pratto et al., 1994). Consequently, people can score high SDO regardless of their own societal power position. These tendencies are of importance, as they imply that high SDO individuals may deny climate change and support anti-environmental actions even if they belong to the social groups that are, or are at risk of being, seriously affected by climate change. It is also important to consider that climate change has been predicted to increase poverty and competition over natural resources (IPCC, 2014). This may lead some individuals to see hierarchies and uneven distribution of climate-related risks even more natural and acceptable, which can make them more accepting of climate injustice.

When it comes to personality underpinnings, SDO has been shown to correlate with empathy (Pratto et al., 1994) and trait-dominance (Grina et al., 2016). That is, individuals who do not empathise with other people and who wish to gain access to resources and powerful positions in society tend to hold positive views regarding group-based hierarchies. These same personality traits could also underpin denial; unconcern for the projected victims of climate change could reduce any sense of urgency about the issue. Indeed, these tendencies are relevant when explaining climate change denial (Jylhä and Akrami, 2015). In particular, (low) empathy predicts climate change denial, and a similar tendency (although not statistically significant) was found for trait dominance. Importantly, SDO mediates both of these relations, and nature dominance mediates the relation between empathy and denial. These results suggest that (low) empathy and trait dominance predispose individuals to accepting group-based hierarchies, which in turn predicts climate change denial.

Concluding remarks

In the light of the psychological research reviewed here, it seems clear that climate change denial is a motivated cognition underpinned by the willingness to maintain the status quo. For example, politically conservative ideology has been consistently shown to be related with climate change denial (i.e. McCright et al., 2016). However, recent research shows that one important explanation for this relation is that conservatives tend to be more accepting when it comes to injustice than liberals (Häkkinen and Akrami, 2014; Jylhä and Akrami, 2015). This finding suggests that climate change denial does not merely reflect a general unwillingness to change, but more importantly seems to include acceptance of unequal distribution of power and risks between different groups of people and between humans and nature. When considering that climate change is mainly caused by the current lifestyle of the wealthy and that it will primarily affect disadvantaged people, future generations and non-human animals (IPCC, 2014), these results seem logical. Climate change denial seems to reflect a motivation to protect and justify the status quo regardless of the negative consequences that it will have on many, both people and animals, today and in the future.

These results are of importance when considering how people could be motivated to support climate change mitigation. The injustice that climate change involves should be better highlighted, as many people may not be aware of this aspect. This information could increase their motivation to change their behaviour in order to lessen their impact on the climate system.

However, when considering the links between climate change denial and empathy, as well as trait dominance, further ways of communication could be considered. Perhaps people who do not empathise with the expected victims of climate change, and who do not wish to jeopardise the resources and power positions that they occupy or wish to occupy, could be reached by other sorts of communication. Indeed, it has been shown that when climate change mitigation is presented as a way to conserve traditional lifestyle, high system-justifying individuals begin to support environmental protection (Feygina et al., 2010). Also, a recent cross-cultural study demonstrated that emphasising co-benefits of addressing climate change, such as economic development or a more moral community, can motivate people to behave in environmentally friendly ways regardless of the level of their belief in climate change (Bain et al., 2015).

It is important to acknowledge that people have multiple concerns regarding climate change. In addition to the negative consequences that the changing climate is causing for people and animals, many worry about the societal changes that could result from mitigation efforts. It would be beneficial to plan both the mitigation policies as well as communication strategies by taking these concerns into account.

References

American Psychological Association. (2009). *Psychology and Global Climate Change: Addressing a Multi-faceted Phenomenon and Set of Challenges. A Report of the American Psychological Association Task Force on the Interface between Psychology and Global Climate Change.* Retrieved from http://www.apa.org/science/about/publications/climate-change.aspx

Bain, P. G., Milfont, T. L., Kashima, Y., Bilewicz, M., Doron, G., Garðarsdóttir, R. B., Gouveia, V. V., Guan Y., Johansson L., Johansson L-O, Pasquali, C., Corral-Verdugo, V., Aragones, J. I., Utsugi, A., Demarque, C., Otto, S., Park, J., Soland, M., Steg, L., González, R., Lebedeva, N., Madsen, O. J., Wagner, C., Akotia, C. S., Kurz, T., Saiz, J. L., Schultz, P. W., Einarsdóttir, G. and Saviolidis, N. M. (2015). How the Co-benefits of Addressing Climate Change Can Motivate Action Around the World. *Nature Climate Change*, 4. Retrieved from http://www.nature.com/nclimate/journal/vaop/ncurrent/full/nclimate2814.html

Budescu, D. V., Broomell, S. and Por, H.-H. (2009). Improving Communication of Uncertainty in the Reports of the Intergovernmental Panel on Climate Change. *Psychological Science*, 20(3), 299-308.

Dhont, K., Hodson, G., Costello, K. and MacInnis, C. C. (2014). Social Dominance Orientation Connects Prejudicial Human-Human and Human-Animal Relations. *Personality and Individual Differences*, 61-62, 105-108.

Duckitt, J. (2001). A Dual Process Cognitive-Motivational Theory of Ideology and Prejudice. *Advances in Experimental Social Psychology*, 33, 41-113.

Dunlap, R. E. and McCright, A. M. (2011). Organized Climate Change Denial. In: J. S. Dryzek, R. B. Norgaard, and D. Schlosberg (eds). *The Oxford Handbook of Climate Change and Society* (pp. 144-160). Oxford: Oxford University Press.

Feldman, L., Myers, T. A., Hmielowski, J. D. and Leiserowitz, A. (2014). The Mutual Reinforcement of Media Selectivity and Effects: Testing the Reinforcing Spirals Framework in the Context of Global Warming. *Journal of Communication*, 64(4), 590-611.

Feygina, I., Jost, J. T. and Goldsmith, R. E. (2010). System Justification, the Denial of Global Warming, and the Possibility of System-Sanctioned Change. *Personality and Social Psychology Bulleting*, 36(3), 326-338.

Grina, J., Bergh, R., Akrami, N., and Sidanius, J. (2016). Political Orientation and Dominance: Are people on the political right more dominant? *Personality and Individual Differences*, 94, 113-117.

Häkkinen, K. and Akrami, N. (2014). Ideology and Climate Change Denial. *Personality and Individual Differences*, 70, 62-65.

IPCC (2014). *Summary for Policymakers. In: Climate Change 2014: Impacts, Adaptation, and Vulnerability. Working Group II Contribution to the Fifth Assessment Report of the Intergovernmental Panel on Climate Change*. Cambridge: Cambridge University Press.

Jackson, L. M., Bitacola, L. M., Janes, L. M. and Esses, V. M. (2013). Intergroup Ideology and Environmental Inequality. *Analyses of Social Issues and Public Policy*, 13(1), 327-346.

Jacques, P. J., Dunlap, R. E. and Freeman, M. (2008). The Organisation of Denial: Conservative Think Tanks and Environmental Scepticism. *Environmental Politics*, 17(3), 349-385.

Jacquet, J., Dietrich, M. and Jost, J. T. (2015). The Ideological Divide and Climate Change Opinion: 'Top-Down' and 'Bottom-Up' Approaches. *Frontiers in Psychology*, 5, 1-6.

Jost, J. T. and Banaji, M. R. (1994). The Role of Stereotyping in System-Justification and the Production of False Consciousness. *British Journal of Social Psychology*, 33(1), 1-27.

Jost, J. T. and Thompson, E. P. (2000). Group-based Dominance and Opposition to Equality as Independent Predictors of Self-Esteem, Ethnocentrism, and Social Policy Attitudes Among African Americans and European Americans. *Journal of Experimental Social Psychology*, 36(3), 209-232.

Jost, J. T., Glaser, J., Kruglanski, A. W. and Sulloway, F. J. (2003). Political Conservatism as Motivated Social Cognition. *Psychological Bulletin*, 129(3), 339-375.

Jost, J. T., Napier, J. L., Thorisdottir, H., Gosling, S. D., Palfai, T. P. and Ostafin, B. (2007). Are Needs to Manage Uncertainty and Threat Associated with Political Conservatism or Ideological Extremity? *Personality and Social Psychology Bulletin*, 33, 989-1007.

Jylhä, K. M. and Akrami, N. (2015). Social Dominance Orientation and Climate Change Denial: The Role of Dominance and System Justification. *Personality and Individual Differences*, 86, 108-111.

Leiserowitz, A., Maibach, E., Roser-Renouf, C., Feinberg, G. and Howe, P. (2013). *Global Warming's Six Americas in September 2012*. New Haven, CT: Yale Project on Climate Change.

McCright, A. M. and Dunlap, R. E. (2011). Cool Dudes: The Denial of Climate Change Among Conservative White Males in the United States. *Global Environmental Change*, 21(4), 1163-1172.

McCright, A. M., Dunlap, R. E., and Marquart-Pyatt, S. T. (2016). Political Ideology and Views about Climate Change in the European Union, *Environmental Politics*, 25(2), 338-358.

Milfont T. L. (2010). Global Warming, Climate Change and Human Psychology. In: V. Corral-Verdugo, C. H. Garcia-Cadena, and M. Frias-Armenta (eds). *Psychological Approaches to Sustainability: Current Trends in Theory, Research and Applications* (pp. 19-42). New York: Nova Science Publishers.

Milfont, T. L., Richter, I., Sibley, C. G., Wilson, M. S. and Fischer, R. (2013). Environmental Consequences of the Desire to Dominate and Be Superior. *Personality and Social Psychology Bulletin*, 39(9), 1127-1138.

Milfont, T. L. and Sibley, C. G. (2014). The Hierarchy Enforcement Hypothesis of Environmental Exploitation: A Social Dominance Perspective. *Journal of Experimental Social Psychology*, 55, 188-193.

Milfont, T. L., Milojev, P., Greaves, L. M. and Sibley, C. G. (2015). Socio-structural and Psychological Foundations of Climate Change Beliefs. *New Zealand Journal of Psychology*, 44, 17-30.

Ojala, M. (2012). Regulating Worry, Promoting Hope: How Do Children, Adolescents, and Young Adults Cope with Climate Change? *International Journal of Environmental & Science Education*, 7(4), 537-561.

Oreskes, N. and Conway, E. M. (2010). *Merchants of Doubt: How aHandful of Scientists Obscured the Truth on Issues from Tobacco Smoke to Global Warming*. New York: Bloomsbury Press.

Poortinga, W., Spence, A., Whitmarsh, L., Capstick, S. and Pidgeon, N. F. (2011). Uncertain Climate: An Investigation into Public Scepticism about Anthropogenic Climate Change. *Global Environmental Change*, 21(3), 1015-1024.

Pratto, F., Sidanius, J., Stallworth, L. M. and Malle, B. F. (1994). Social Dominance Orientation: A Personality Variable Predicting Social and Political Attitudes. *Journal of Personality and Social Psychology*, 67(4), 741-763.

Schlosberg, D. (2013). Theorising Environmental Justice: The Expanding Sphere of a Discourse. *Environmental Politics*, 22(1), 37-55.

Spence, A., Poortinga, W. and Pidgeon, N. (2012). The Psychological Distance of Climate Change. *Risk Analysis*, 32(6), 957-972.

Wilson, M. S. and Sibley, C. G. (2013). Social Dominance Orientation and and Right-Wing Authoritarianism: Additive and Interactive Effects on Political Conservatism. Political Psychology, 34(2), 277-284.

SECTION II:

ASSESSMENTS – WHICH WAY TO FOLLOW?

6

Transversal Environmental Policies

GUSTAVO SOSA-NUNEZ
INSTITUTO MORA, MEXICO

Modern times have seen the environment degraded due to careless production and consumption. Attempting to overcome this, nation-states have developed environmental policies according to their own perspectives, interests and geopolitical strategies. Many of them are traditionally viewed as 'inherently regulatory', dominated by national governments stipulating in law-specific standards (Jordan et al., 2013: 168). Of which, some are characterised by a prevalent focus on the 'here and now'; meaning that we often react to problems only when they affect our daily lives (Rudel, 2013: 2).

The amelioration of pollution and the preservation and protection of natural resources are two issues that remark the importance of regulating interactions between societies and the environment. Being made at both the national and international levels, policies present different features. On one hand, environmental management at national level is characterised by different domestic policy-making processes, economic and ecological conditions, sociocultural values, levels of activism, as well as land-use and natural resource regimes (Healy et al., 2014). On the other hand, the international level promotes an understanding of the implications of environmental problems that do not recognise political boundaries, implying the potential development of global, international and regional policies. This stance suggests the consideration of additional variables, such as the type of interactions that nation-states sustain at global forums, political parties in power, and national interests, to cite a few.

The varied features mean that a multi-field approach is suitable to understand the environmental policy landscape. There are different interpretations for 'field'. Boasson (2014: 27) enlists DiMaggio and Powell (1983), Bourdieu and Wacquant (1992), Greenwood et al., (2002) and Scott (2008). Common ground corresponds to relationships among individuals and societies, or between sections of societies, in their aggregate, around specific industries, or merely as part of institutional life.

Diversity of related policy areas, instruments, institutions and actors shape policy-making towards what they understand and identify as best for their natural surroundings. Despite of the acknowledgement of common shared environmental values; political culture, identity, idiosyncrasies, and interests make national and local environmental objectives to differ. In this sense, variables like personal ability, motivation, corruption and nepotism should be of mandatory consideration, as they can often shape environmental values, identities, and national interests. To do this, it is important to identify the roles that humans – politicians, businessmen, social actors, etc. – and institutions – in subnational, national and regional arenas – play.

The multi-field approach helps to explain the reason and the manner in which political actors behave on policy-related issues. It also allows setting up the role that other political and non-political actors play in environmental policy-making processes. Furthermore, this approach uses the concept of multi-level governance, which refers to decision-making interlinked at different levels (international, national, subnational and local) and geographical areas over a specific issue. The relationship 'field-level' relies on the use of policy systems that is made across multiple hierarchical levels (Boasson, 2015: 26).

Considering that policies dealing with environmental matters should be developed in accordance to other seemingly unrelated policies, the multi-field approach provides the opportunity to examine diverse aspects. First, it allows observing the links it has with broader sets of policies. Second, it assists to explain the policy interaction that national governments have between and within them, especially in cases when transboundary environmental problems occur. In this context, this chapter aims to offer an insight about the role environmental policies play in overall policy frameworks. The transversal nature of these policies has long been acknowledged. However, this does not mean that environmental policy is accurately related – let alone integrated – to a wider policy framework. There are cases in which the link is subtly established. In some others, the relationship is clearly and properly set up. Notwithstanding, the importance of environmental policies is not equally recognised. At times, they play a central role. In other cases they are peripheral to policy developments.

This background allows setting the present contribution as follows. The second section presents divergent approaches to include the environment in policy frameworks – administrative rationalism, democratic pragmatism and economic rationalism – and the importance of environmental policies. The third section distinguishes the role that environmental policies have in broader policy frameworks. In this sense, explanations about the intrinsic role that environmental policies have with other policies are presented. For this, industry, security, science, climate change and urban planning are used as examples. The fourth section then aims to identify the adequate conceptualisation of environmental policies. It questions whether transboundary cooperation or international governance better explain the transversal approach that environmental policies have. Lastly, conclusions observing if environmental policies are central or peripheral are shown.

Divergent policy approaches and the importance of environmental policy instruments

Each nation-state has its own environmental policy style. Within its realm, different ministries, even political parties, 'formulate environmental policies as part of their ordinary activities, regardless of whether they believe in them' (Buzan et al., 1998: 73). They treat environmental problems as 'tractable within the basic framework of the political economy of industrial society' (Dryzek, 2013: 73).

In order to dissect them, three varying approaches have been recognised. One is administrative rationalism, which encompasses the dominant governmental response to environmental problems, emphasising the role of experts over the citizenry. Institutions identified in this approach are pollution control agencies that exist at international, national and subnational levels. However, there is no global perspective being identified, which means that expertise and research can be influenced or driven towards a preferred perspective or ideology. Right-wing politicians have even claimed that scientific neutrality is effectively impossible (Dryzek, 2013). The second approach is democratic pragmatism. Thought of as a response to the shortcomings found in administrative rationalism, this approach aims to make administration more responsive and flexible according to the circumstances that exist at a given time period (Fiorino, 2004). For this to happen, democratisation of environmental administration is necessary; which can happen through public consultation, alternative dispute resolution, policy dialogue, lay citizens' deliberation, public enquiries or right-to-know legislation. Of course, any of these types implies – or intends –widening the scope of participating actors. The third approach, economic rationalism, offers a way for market mechanisms to reach objectives of public interest. Governments should play a peripheral role. Their participation would relate to setting up basic market rules, with the potential implication of natural resources privatisation. Developing markets in environmental goods would provide a further pathway of action, one regarded as environment protective by supporters (Dryzek, 2013: 100–124).

Whether these approaches are successful depends on the context and the extent to which such policies are formulated and implemented. This also relates to differences within and between nation-states in attitudes and behaviours towards the environment, which can be either benign or self-destructing (Watts, 1999: 266).

The politicisation of the environment may relate to the knowledge governments and societies have about the natural world and the interaction that humans have with it. However, these actors may not share the same perspective nor have a similar position of power. The citizenry tends to be more environmentally minded than its government; despite the environment not being located high in the list of concerns that many sectors have, especially in developing countries. Income, economic development, health and security are four areas generally located above environmental concerns.

Environmental policies are carried out through policy instruments, understood as the numerous practices at the disposal of governments to implement and reach their policy objectives (Howlett, 1991: 2). They assist to clarify the relationship between government – the state-led governing that relies on laws – and governance – used with horizontal forms of societal self-coordination (Jordan et al., 2005); although they are influenced by policy-makers' goals, outlooks and philosophies defining national interests (Hall, 1993). Jordan et al. (2013) comment that environmental policy instruments are aiming at an interdisciplinary holistic approach that takes into account political processes and contextual factors – such as voting rules, power of industry, dominant ideas and policy paradigms – that shape designs, calibrations and usage of such instruments. Market-based instruments, including emissions trading and eco-taxes, are some of the most important. Some others are used to provide information – such as eco-labels and management systems – or are set up through voluntary agreements, like clean development mechanisms. The ultimate instrument becomes the regulation, as its mandatory status infers reaching policy objectives. For this purpose, implementation programmes are developed, although they have not been sufficient. In many cases, failure to adequately implement environmental programmes responds to features such as corruption, lack of expertise and technical unviability in remote places. In some cases, secondary programmes are formulated to assist in the implementation of main programmes; but this spillover of programmes does not ensure that implementation takes place.

Interdisciplinary role of environmental policies

Intended to preserve and protect the environment, policies in this area are interdisciplinary – incorporating understandings of the natural and social sciences as a means to understand routes forward. Nonetheless, they are not always branded and developed as such, as widening the scope of a given policy can pose serious limitations to implementing institutions. The interdisciplinary status should be regarded as a challenge (Salter and Hearn, 1996), one that allows more freedom and creativity to work across different types of experience and fields of knowledge (Hackett and Rhoten, 2009).

One area in which environmental policies participate – or should do so – is the industrial sector. Both industry and government 'evolve and function in accordance with governmental regulations' (Boasson, 2015: 12), with the possibility to develop shared worldviews and preferences on the environment. In this sense, there are current trends showing an interest to reduce carbon emissions while saving money in the process. For this to happen, initial investment in reduction projects is required. Acquisition of greener company vehicles, more efficient production gear and processes, use of recycled materials, reduction of energy consumption and methane emissions cuts are, among others, diverse actions made under the auspices of certain industrial programmes set up as a result of public policies aimed at improving environmental conditions. However, there are many key sectors that pose environmental challenges due to their core processes, like plastic, paper, automotive, agricultural, livestock, energy-producing and energy-intensive industries.

A further policy area found in this context is security. Scarcity, detrimental effects of resource use and environmental destruction due to expansive economic activity can all lead to conflict (Dalby, 2014: 230). Potential threats to a state, its population, or its natural resources can update security policy. A perspective from International Relations studies suggest that a matter becomes an issue of security or emergency when securitising actors affirm that something constitutes a threat to an object that needs to survive and hence should be dealt with immediately (Floyd, 2010). In turn, the objective becomes the 'desecuritisation' of the threat at hand, understood as the process by which securitisation is reversed and the threat disappears, leaving the issue out of emergency mode (Buzan et al., 1998). Given the fluctuations of environmental threats – like an increase in floods, high intensity hurricanes, wildfires, deforestation and droughts – and the lack of true commitment to act as international community, desecuritising the environment is not suitable for the time being. In fact, a broader range of areas may be securitised in the near future. Freshwater supply for human consumption fits in this case. There is a trend showing an increase in water shortage in many regions of the world. Two-thirds of the world's population may face this problem by 2025, with subsequent implications for ecosystems and wetlands (WWF, 2015). Hence, competition over declining natural resources, like freshwater, may lead to conflict between and within states. For this reason, an international policy approach – consistent with national lines – linking the environment and security needs to be developed. In this sense, an Environment and Security Initiative is under development, assisting national governments, as well as their local communities, to identify common solutions and develop work programmes and project portfolios (Environment and Security Initiative, 2015). However, this initiative is not planned on a global scale, focusing instead on four regions at present: Central Asia, South Caucasus, Eastern and South East Europe.

Science is another area where policy needs to interact with the environment. Its relevance includes different perspectives. In security terms, the scientific perspective of a policy can relate to the authoritative assessment of threat for securitising or desecuritising moves (Buzan et al., 1998: 72). Science is essential for knowledge-sharing, which can assist to develop policy frameworks aimed at protecting the environment, foment sustainability, and promote social welfare. This can be made through adequate international cooperation strategies destined for such purposes. Research findings are the basis by which policy is – or should be – formulated, developed and implemented. In fact, referencing science to make people understand about the importance and implications of mishandling the environment is becoming the norm. However, science can also be used to develop counter-arguments about the context in which the environment is found. For example, it can be used to deny the existence of climate change – as it has been used to claim its existence.

Energy is a sector that is intrinsically related to the environment. Fossil fuels remain the most used energy source despite of common knowledge about its negative implications for local eco-systems and climate change. Indeed, there is a growing industry for renewable energy. Eolic (wind), geothermal (heat from Earth), biomass (biological material), hydroelectric, ocean, and solar sources are slowly becoming operational; although it will take them some time to overcome hydrocarbon production – if we are to see it. Nuclear energy is also in the frame, although it deserves special consideration due to its potential catastrophic environmental consequences in case of accidents or careless use. All these types of sources are getting included in national energy legislations, some of which show signs of a true commitment to a shift of energy origin, such as those of Austria, Germany, Denmark and Sweden. All four are already setting deadlines for the change to take place (go100percent.org, 2015). Some other countries refer to these mechanisms but are slow to move in such a direction, as is the case with China (Pedong et al., 2009) and Mexico – the latter characterised by an oil industry representing an important source of revenue (Sosa-Nunez, 2015). A third group of countries would not contemplate renewable energy. This is the case with members of the Organisation of the Petroleum Exporting Countries (OPEC), who insist that fossil fuels will still contribute 82 per cent of the world's energy supply in 2035 (OPEC, 2012). Their position is understandable, as a change of direction would be against their national interests (Assis, 2014).

A further topic is climate change. The degrading trend of the environment has led the international community to join their efforts to develop a common framework to combat it. Understanding that climate change derives from humankind's overproduction, which releases overwhelming amounts of carbon dioxide and other polluting emissions; policymakers attempt to develop policies on modifying the population's behaviour and consumption patterns towards a more sustainable approach. For this, production processes also need to change, implying a potential overhaul of existing industrial policies. Obviously, reticent stances from many sectors of the world society are likely; but the present environmental situation leaves no room for further delay given the delay embedded in the setting of long-term deadlines to accomplish what are already overdue objectives.

The world's population is expanding faster than ever, meaning an increase of urban areas – in both size and number. Urban planning is yet another area in which environmental policies should play a key role, looking at adequate land use, electric public transport, cycling lanes, distributional networks to reduce traffic, adequate waste disposal and even the location of recycling bins. Actions in many of these areas would not only improve the environmental conditions of the cities in which they are implemented; they would also lead to significant savings, regardless of the size of the cities or whether they are in developing or developed countries (where these policies are not fully implemented). The 'smart cities' approach could help to reduce energy and water use and waste generation. At the same time, this approach should have an emphatic environmental perspective that combines infrastructure improvements, connectivity, social and societal development, as well as citizens' wellbeing. Only then will cities be able to meet their environmental targets.

Industry, security, science, climate change and urban planning are just five of many policy areas that are intrinsically related to environmental policies. Other areas seem to approach the environment in slightly different ways. This is the case with the economy. Findings suggest that economic activity damages the environment; but continued economic growth – specifically, per capita income increases – eventually reverses this trend. (Stern, 2004) In fact, there are claims that encouraging economic growth is the best environmental policy option. However, this argument only stands as long as pollutant impacts are immediate and local: intergenerational impacts are not contemplated (Raymond, 2004: 327), neither is the long-term nature of climate change.

These areas provide sufficient information about the transversal nature of environmental policies. An environmentally conscious approach is easily identified with science and climate change matters, since these look to preserve natural resources or to ameliorate negative effects that we – humans – have had on our natural surroundings and the planet. At the other end of the scale, perspectives on environmental security place societies' security concerns before awareness of the environmental situation. The industry and urban planning examples are located somewhere in the middle. They have willingly – or reluctantly – embraced environmental policies, but improvements are required. For this to happen, the role of national and local governments is of utmost importance, as they set the rules by which we should play.

Transboundary cooperation or international governance?

The current environmental context – characterised by modifying worldwide climate patterns – has led some to argue that an 'environmental crisis' exists, making global collective action a necessity (O'Neill, 2009). Problems such as biodiversity loss, deforestation, waste management, ozone depletion, atmospheric pollution and ocean acidification mean that collaborative counteractive efforts are mandatory.

The type of action depends on the type of problem to solve. A global response is needed to tackle climate change, while a regional approach is more suitable for improving transboundary air pollution or reforestation. Moreover, collective action has to take place vertically across multiple levels of government and horizontally across multiple arenas involving public and private actors. The interdependence of actors means that cooperation is a necessary action when governing across multiple arenas. As it includes traditional modes of government and non-hierarchical modes of guidance (Héritier, 2002: 1), the resulting political steering is conceptualised as governance.

The context in which governance takes place indicates the existence or absence of legislative decision-making. Whether it is one way or another depends on the extent of the participation by public and private actors. Some forms of governance – diffusion, learning, knowledge standardisation, repetition, persuasion – can avoid regulatory requirements by promoting voluntary actions, although these can be influenced by rapidly changing economic, social and technological contexts. These in turn have implications for the type of policy outcomes.

With regard to the governmental perspective, collaborative efforts that nation-states make depend on their inner preferences, interests and needs. Vicinity plays a part when setting up environmental policies of a certain kind. This is the case for northern member-states of the European Union (EU), whereby Denmark, Sweden, the Netherlands and Germany share high levels of domestic commitment and a precautionary approach to policy at international level, which is replicated at EU level (Watts, 1999). Oppositely, there are cases where political boundaries define contrasting positions. The United States of America and Mexico exemplify this, as the former has stricter environmental regulations than the latter (although this is slowly changing – Mexico is catching up as a result of trade and environmental agreements that are reshaping their relationship).

Conclusions: Are environmental policies core or peripheral?

There is an increasing tendency to get environmental policies involved in broader, seemingly unrelated policies; however, more needs to be done if we are to improve the environmental conditions that characterise our time. In this sense, the interdisciplinary spectrum of environmental policies includes not only socio-economic components but also physical, biological, mathematical and engineering ones. The environment is also considered in newer research, health and education programmes. This is desirable, since they should 'be guided by a sense of the greater public good and a reliable moral compass' (Hackett and Rhoten, 2009: 409).

Natural sciences and technology explain the physical processes that we face but not the arguments behind choosing not to confront them (Regan, 2015: 4). Whether it is through incentives or coercion, we need to modify our consumption patterns while also learning how best to interact appropriately with the environment. Policies in such a direction should prove fruitful.

Perhaps the way forward is about formulating and implementing policies focusing on zero economic and population growth. A conservative approach would be necessary. However, this perspective attracts a lot of criticism from those stating that we should protect the environment and the planet as long as we, humans, are not affected in economic terms. The right to become – or remain – developed societies is claimed as central to this argument. However, considering the levels of environmental degradation that we have caused, the only way to reverse trends implies establishing commitments and sacrifices from all societies and all sectors. We should behave as a single race: humankind. Despite its promotion, moderate environmentalism ¬– advocating economic growth while reaching limits of natural resource consumption – is no longer suitable, or at least not until ecosystems recover from overuse and pollution.

The role of environmental policies in overall policy frameworks has been increasing, although there is a long way to go. Their presence can assist in establishing an environmental policy reform in which the different participating – or willing to participate – actors promote the development of a strict approach to environmental protection and conservation. The relevance of each individual's actions – at local levels – should also contribute to international and global efforts to tackle environmental problems. If this occurs, implementation may be feasible and smooth.

References

Assis, C. (2014). *Renewable-Energy Stocks Hobbled by OPEC*. Marketwatch. 28 November. Retrieved 17 September 2015, from http://www.marketwatch. com/story/renewable-energy-stocks-hobbled-by-opec-2014-11-28

Boasson, E. L. (2015). *National Climate Policy: A Multi-filed Approach*. Abingdon: Routledge.

Bourdieu, P. and Wacquant, L. J. D. (1992). *An Invitation to Reflexive Sociology*. Cambridge: Polity Press.

Buzan, B., Wæver, O. and De Wilde, J. (1998). *Security: A New Framework for Analysis*. Boulder, CO: Lynne Rienner.

Dalby, S. (2014). Security. In C. Death (ed.). *Critical Environmental Politics* (pp. 229-237). Abingdon: Routledge.

DiMaggio, P. J. and Powell, W. W. (1983). The Iron Cage Revisited: Isomorphism and Collective Rationality in Organizational Fields. *American Sociological Review*, 48(2), 147-160.

Dryzek, J. S. (2013). *The Politics of the Earth: Environmental Discourses*. 3rd.Third Edition. Oxford: Oxford University Press

Environment and Security Initiative (2015). *About Us*. Retrieved from http://www.envsec.org/index.php?option=com_content&view=article&id=60&Itemid=176&lang=en

Fiorino, D. J. (2004). Flexibility. In: R. F. Durant, D. J. Fiorino and R. O'Leary (eds). *Environmental Governance Reconsidered: Challenges, Choices, and Opportunities* (pp. 393-425). Cambridge, MA: MIT Press.

Floyd, R. (2010). *Security and the Environment: Securitisation Theory and US Environmental Security Policy*. Cambridge: Cambridge University Press.

Go100percent.org (2015). *Projects in Europe*. Retrieved from http://www.go100percent.org/cms/index.php?id=19

Greenwood, R., Suddaby, R. and Hinnings, C. R. (2002). Theorizing change: the role of professional associations in the transformation of institutional fields. *Academy of Management Journal*, 45(1), 58-80.

Hackett, E. J. and Rhoten, D. R. (2009). The Snowbird Charrette: Integrative Interdisciplinary Collaboration in Environmental research Design. *Minerva*, 47(4), 407-440.

Hall, P. A. (1993). Policy Paradigms, Social Learning and the State. *Comparative Politics*, 25(3), 275-296.

Healy, R. G., VanNijnatten, D. L. and López-Vallejo, M. (2014). *Environmental Policy in North America: Approaches, Capacity, and the Management of Transboundary Issues*. Toronto: University of Toronto Press.

Héritier, A. (2002). *New Modes of Governance in Europe: Policy Making without Legislating?* Political Science Series 81. Vienna: Institute for Advance Studies.

Howlett, M. (1991). Policy Instruments, Policy Styles and Policy Implementation. *Policy Studies Journal*, 19(2), 1-21.

Jordan, A., Wurzel, R. K. W. and Zito, A. R. (2005). The Rise of 'New' Policy Instruments in Comparative Perspective. *Political Studies*, 53(3), 477-496.

Jordan, A., Wurzel, R. K. W. and Zito, A. R. (2013). Still the Century of 'New' Environmental Policy Instruments? Exploring Patterns of Innovation and Continuity. *Environmental Politics*, 22(1), 155-173.

O'Neill, K. (2009). *The Environment and International Relations*. Cambridge: Cambridge University Press.

OPEC (2012). *OPEC's View on the Global Energy Scene*, 28 September. Retrieved from http://www.opec.org/opec_web/en/2426.htm

Peidong, Z., Yanli, Y., Jin, S., Yonghong, Z., Lisheng, W. and Xinrong, L. (2009). Opportunities and Challenges for Renewable Energy Policy in China. *Renewable and Sustainable Energy Reviews*, 13, 439-449.

Raymond, L. (2004). Economic Growth as Environmental Policy? Reconsidering the Environmental Kuznets Curve. *Journal of Public Policy*, 24(3), 327-348.

Regan, P. M. (2015). *The Politics of Global Climate Change*. Boulder, CO: Paradigm Publishers.

Rudel, T. K. (2013). *Defensive Environmentalists and the Dynamics of Global Reform*. Cambridge: Cambridge University Press.

Salter, L. and Hearn, A. (1996). *Outside the Lines: Issues in Interdisciplinary Research*. Montreal: McGill-Queen's University Press.

Scott, W. R. (2008). *Institutions and Organizations: Ideas and Interests*. 3rd Edition. Thousand Oaks, CA: Sage.

Sosa-Nunez, G. S. (2015). Climate Change Policy and Energy Reform: An Assessment of Mexico's Foreign Policy. *Latin American Policy*, 6(2), 240-254.

Stern, D. I. (2004). The Rise and Fall of the Environmental Kuznets Curve. *World Development*, 32(8), 1419-1439.

Watts, N. (1999). The Hare and the Tortoise: Dead in the Heat? Cross-National Differences and Knowledge Gaps in Environmental Policy. *Politics and the Life Sciences*, 18(2), 266-268.

WWF (2015). *Threats: Water Scarcity*. Retrieved from http://www.worldwildlife.org/threats/water-scarcity

7

Environmental Conflict: A Misnomer?

ED ATKINS
UNIVERSITY OF BRISTOL, UK

With the decline of the Cold War and the traditional concerns that the period embodied, academics and policymakers began to redefine what security means – with a greater focus on the environment, degradation and scarcity as a route to conflict. Over the past 25 years a new field of research has developed, in which the interactions between the environment and conflict are examined. Within this field, a paradigmatic causal chain has emerged: that population growth, by increasing consumption and production, will cause environmental deterioration and scarcity, thus exacerbating competition and creating conflict.

As Le Billon (2012: 9) states, 'the idea that wars are associated with resources is probably as old as war itself'. Ever since Thomas Malthus wrote his 'Essay on the Principle of Population' in 1798, the connections between environmental problems, competition over resources and violent conflict have captured the minds of many. This appears simple: the environment is a key driver of human civilisation, and when security and resilience cannot be found, warfare is often what we turn to to achieve our aims and secure resources (Westing, 1986). Recent findings at the Neolithic massacre sites of Talheim (Germany) and Asparn and Schletz (Austria) have provided further evidence of this turn to conflict – with evidence of ancient massacres occurring as the result of completion between the first Central European farmers (Meyer et al., 2015). Within the spectre of climate change, the environmental-conflict thesis asserts that such a process will result in increased rates of conflict over scarce resources, degraded environments and the environmental refugees that will likely flee affected areas (Hsiang et al., 2013).

However, this assertion possesses a number of flaws that alter its credibility and wider applicability. The neglect of the fundamental differences between violence as a consequence of environmental degradation and the economic nature of non-renewable resources as a driver of armed conflict has resulted in the conflation of struggles of different natures and involving divergent motivations. Additionally, ignorance of alternative variables provides another important limitation. Historical evidence suggests that the environment acts as just one component of a larger, complex web of causality, interacting with a number of alternative variables of both political and social composition (Salehyan, 2008). As a consequence, it will be posited in this chapter that no conflict can be exclusively environmentally *driven*; rather, that violence has only been environmentally *induced*, with ecological factors acting with a number of other factors to create a structure that allows for an escalation into conflict.

First: a caveat. It is important to differentiate between the concepts of conflict and insecurity. Although the terms are regularly treated as synonymous, due to the mutually constitutive causal chain between them, conflict is an empirical phenomenon that can be observed, while the concept of insecurity is subjective and socially driven (Dabelko et al., 2000). Consequentially, it is necessary to examine cases of violent conflict within the literature rather than examples of insecurity linked to environmental degradation and scarcity. Several opponents of the thesis criticise the framing of the environment within a narrative of security, perceiving it as a means to legitimise new areas of military activity. References to security often result in the construction of popular approval for prescribed action, due to the terms' resonation with popular desires to remain unthreatened (Dalby, 2002). Peter Haas (2002) has explored the creation of such securitised discourse surrounding environmental conflict, asserting that such practices are regularly invoked as a means to justify pre-existing state goals and do not constitute a window into state resilience. A number of authors have echoed these assertions, stating that such declarations are often made by certain interests, as a means to legitimise wider political agendas and motives (see Barnett, 2000; Verhoeven, 2011; Selby, 2014).

Separating the environment from the economy

History provides many examples of the role that the environment plays in the creation of armed conflict, fundamentally characterised as violent: from the Peloponnesian War to Alexander the Great's empire. Westing (1986) identified 12 international conflicts over natural resources, both renewable and finite, in the period between 1914 and 1980, with examples including: World War One (the result of German desire for oil); the operational role of *lebensraum* in the German activity that resulted in World War Two – with similar motivations for Japanese expansionism; and the role of control over the River Jordan as a cause of the Arab-Israeli War of 1967. A territory's possession of resources has also provided motivations for the aggressive behaviour of colonial powers when faced with independence struggles, as illustrated by the examples of Algeria (1954–62), Belgian activities in Congo (1960–64), and the Western Sahara Revolt (1976 onwards).

This strategic control of resources, and the financial wealth that such control brings, has often motivated internal armed movements (Klare, 2001). The United Nations Environment Programme (UNEP) (2009) found that 40 per cent of all intra-state conflicts are linked to the appropriation or control of natural resources. Charles Taylor's violent appropriation of timber and mineral rights, in both Liberia and Sierra Leone, was conducted as a means to finance the rebel forces of the National Patriotic Front of Liberia (Klare, 2001). Additionally, the National Union for the Total Independence of Angola's sale of blood diamonds not only allowed for the financing of war but also resulted in the procurement of political support from the nations of Togo and Burkina Faso (Le Billion, 2001). In both cases, the presence of natural resources led to the commercialisation of armed conflict. Collier and Hoeffler (2004) found that an abundance of resources heightens the risk of violent conflict, due to this prize of control. Within this understanding it is the profit of these resources that allow the continuation of armed movements in Aceh, Indonesia; Biafra, Nigeria; and across the globe (Bannon and Collier, 2003). Such sub-national conflicts provide the thesis with its most important evidence.

Importantly, a widely cited example of resource conflicts involves the presence of oil. Lujala (2010) pursued the Collier-Hoeffler findings further, finding that the presence of oil and gas reserves in arenas of violence has an important effect on the duration of conflict, with on-shore oil production often increasing the risk of potential conflict. This is symbolic of the increased entrenchment of oil within patterns of foreign policy and, with it, frequent associations between the resource and conflict. In his 1986 work, Westing pointed to a number of examples of conflicts over oil – from the 1932–35 Chaco War, between Paraguay and Bolivia; to the Paracel (His-sha) Islands clash of 1974, in which China routed Vietnamese forces to reclaim this group of islands in the South China Sea. However, these episodes were over non-renewable resources. As a consequence, such conflicts are tied not to the environment *per se* but rather to the economic and commercial nature of the resources themselves. This represents an important flaw in the environmental-conflict thesis: the conflation of non-renewable resources that possess a quantifiable economic value, such as oil and gas, with important ecological issues, such as degradation and climate change (Barnett, 2000).

It is not the scarcity of resources that drive these conflicts, but the greed associated with their abundance. Although Saddam Hussein invaded Kuwait as a means to control the oil fields of his neighbour, he did so as a vehicle to boost Iraq's flailing economy (Klare, 2001). As such, the conflict was not motivated by the resource but by its commercial value. If oil prices, and global demand, had been low, the conflict would have been futile and would not have occurred. The conflicts that took place in Liberia and Angola were not explicitly over control of resources but rather the affluence this would provide, enabling investment in the continuation of the war effort. Rebel groups were not fighting for the control of the resource but targetted them as a means of sponsorship (De Koning, 2008).

It can be asserted that such conflicts were not environmental conflicts *per se* but rather traditional conflicts related to the commercial value of the environment (Libiszewski, 1992). It is society's interactions with this 'material' that bestow it with an economic value, and it is this importance that drives conflicts over the resource. This hybrid socio-nature of economic resources is driven by production: the transformation of oil from fossilised organic material to the petrochemicals that are central to our lives. While nature creates these materials, it is societal reliance that converts them into conflict-worthy materials.

The economic and strategic importance of oil and other non-renewable resource is indisputable. Yet the globalised character of international commerce has resulted in many nations ceasing to perceive resource dependency as a threat to autonomy or survival (Deudney, 1990). This interdependence has resulted in the decreased likelihood of inter-state conflict over control of resources, due to the price shocks these actions could propel across the system and the increasingly technological developments (Lipschutz and Holdren, 1990). Such dynamics are well illustrated by the 1973 oil crisis (Dabelko and Dabelko, 1993).Although the move by the Organisation of Arab Petroleum Exporting Countries (OAPEC) to restrict exports resulted in record price rises and the transformation of the international sphere, thus illustrating the economic relevance of resources, it did not result in international violent conflict. Furthermore, Le Billon (2001) has stated that the spectre of resource scarcity has resulted in the escalation of socioeconomic innovation and economic diversification – with the market mechanisms of contemporary capitalism creating an important impediment to conflict. In Botswana and Norway, minerals and oil, respectively, have been mobilised to ensure peaceful development rather than violent confrontation (Le Billon, 2001). Furthermore, in many cases potential scarcity has resulted in increased inter-state cooperation due to the shared interest in continued supply. The continued sanctity of the 1960 Indus Waters Treaty, between Pakistan and India, is an important example, with the spirit of cooperation over water resources enduring despite increased political tensions between the two nations (Wolf, 1998).

Defining environmental conflict

Many have defined an 'environmental conflict' as one resulting from degradation caused by human activity or mismanagement, rather than the finite nature of the resources exploited (Dokken and Graeger, 1995; Benjaminsen et al., 2012). Notably, this definition will include renewable resources such as agricultural land, forests, water and fishing stocks – all of which depend on sustainable usage to ensure their continuation. The operating role of the ecological factors as a cause of conflict has been relatively neglected, with much of the academic focus being on the environment as an economic resource (WBGU, 2007).

It is important to note that such conflicts predominantly occur on an intra-state basis, rather than between two nations. International conflict over environmental factors remain unlikely – whether due to the robust nature of the world trade system and dynamics of supply and demand or to the spread of small arms transforming the notion of traditional conflict (Deudney, 1990). An important example can be found in the assertions of *water wars*. Although the management of rivers is often complicated by their crossing of territorial boundaries and nations dependent on water from beyond their borders (Egypt, Hungary and Mauritania all rely on international watercourses for 90 per cent of their water), an international conflict exclusively over possession of and access to a shared water source is still to occur. The reasons for this are simply, as Wolf (1998: 251) states, 'War over water seems neither strategically rational, hydrographically effective, nor economically viable.' At the international level, the costs outweigh the benefits and cooperation is sought before conflict occurs.

Environmental problems are not first felt at the international level but provide challenges to communities within the nation-state. Barnett and Adger (2007) suggested four key factors in the creation of environmental conflict: poverty, vulnerable livelihoods, migration and weak state institutions – all problems that are present at the local level. The German Advisory Council on Global Change (WBGU, 2007) found 73 examples of such sub-national ecological conflicts, primarily linked to natural disasters, between 1980 and 2005. These cases included violent regime change (1954 – Hurricane Hazel in Haiti), civil war and secession (1970 – the Bhola cyclone in East Pakistan/modern-day Bangladesh) and anarchy and looting (2005 – post-Katrina New Orleans) (WBGU, 2007).

All such conflicts took place within wider causal structures. The violence in both Haiti and Bangladesh occurred due to the role that the climactic shocks took in exacerbating existing tensions (WBGU, 2007). Environmental factors often interact with the visible drivers of ethnic tensions, political marginalisation and poor governance to create a causal framework that allows degradation to affect livelihoods, interests and capital – which, in turn, lead to conflict. The contemporary occurrence of violent unrest over food and water prices provides an example of popular outbursts of dissent and violence as the result of environmental factors. During the 2008 food price crisis, 61 nations experienced some form of unrest in protest at the conditions; 38 of these were of a violent nature (Castles, 2002). Notably, all such conflicts take place on a localised, sub-national scale, with violence rarely straddling borders – illustrating the importance of contextual factors. The links between environmental stress and conflict are indirectly constructed, with a structural vulnerability necessary for the transition from poor harvests, caused by climactic factors, to food shortages and, thus, price inflation. Their occurrence is directly linked to the presence of a political vacuum in which the government either cannot provide stability and representation or is engaged in corruption and rent-seeking (Benjaminsen et al., 2012).

Looking beyond nature

All conflicts have particular geographies, in which a number of factors interact. Economic stability, regime type, demography, patterns of consumption, historical consciousness and power dynamics all play a role in the construction of this milieu (WBGU, 2007). Such factors affect the vulnerability of populations, institutions and ecosystems to degradation and environmental change (Evans, 2010; Gleditsch, 1998). Exposure to climatic shocks is likely to exacerbate existing societal tensions, such as poverty and exclusion, thus creating a driver for conflict in many regions (Levy, 1995). The actual risk of violence, as a consequence of environmental change, is dependent on this vulnerability (Evans, 2010). In Sudan, the Darfur conflict was the result of poverty and falling incomes, compounded by population growth and environmental pressures – ecological issues were not standalone (Benjaminsen et al., 2012). These factors are closely intertwined with those of politics and economy, resulting in a causative labyrinth – the presence of which results in assertions of exclusively environmentally driven conflicts becoming misguided and speculative. Sadly, this wider causality has been neglected in much of the research into the topic, representing an important methodological weakness (Hauge and Ellingsen, 1998). This can be seen in Selby's (2014) review of 33 quantitative environmental conflict studies, which found no real consensus on the linkages between the environment and conflict. As Selby (2014: 834) states, 'One study finds that low rainfall is associated with increased conflict in Sub-Saharan Africa; two others find instead that it is high rainfall that is linked to increased conflict in Africa; two studies find that high rainfall is, to the contrary, associated with reduced conflict in Africa; a further two have found that precipitation extremes, of either sign, are associated with increased African conflict; one study finds that drought has historically been associated with conflict in Europe; and five studies find no meaningful correlations between rainfall and conflict.' This lack of consensus is driven by the neglect of important social, political, economic and cultural factors in the creation of conflict situations, and a focus on the environment as a primary cause.

In these cases, it is human factors rather than the environment that lie behind societal breakdown and violence. Although the unprecedented population growth in Rwanda, coupled with traditional competition between the Hutu cultivator and Tutsi pastoralists, can be considered a primary causal factor in the 1994 genocide (Dabelko et al., 2000), causality can also be attributed to the Belgian colonial legacy that encouraged such division, alongside other structural causes of conflict (Castles, 2002). Colonial legacies have been blamed for the ecological grievances of many, with Moorehead (1992) stating that violent conflict in Mali was strongly linked to the land tenure system, a product of French colonial rule, rather than the scarcity of fertile land.

Similarly, state resilience in the face of environmental problems plays a pivotal role in the onset of violence, with climate change reducing capacity to sustain the livelihoods of the population (Barnett and Adger, 2007). Hauge and Ellingsen (1998) found levels of economic development and regime type to be powerful predictors of the outbreak of conflict, with degradation acting as contributory cause rather than driver. Consequently, it has been predicted that the perils of environmental conflict are more likely to occur in less developed, poorly governed states (Smith and Vivenkananda, 2007). For example, although deforestation and the resultant soil degradation led to the collapse of the state of Haiti in the 1980s and 1990s, it was the successive repressive regimes of Port-au-Prince that exacerbated such issues by failing to ensure human security within the nation (Elliott, 1998). These competing variables also result in the absence of conflict in other states. Although desertification and food insecurity were considered of causal importance in the civil war in Darfur, such factors have not led to conflicts in other states (Salehyan, 2008).

Thomas Homer-Dixon (1994) explored the potential for violence at a local level as a result of environmental degradation and resource scarcity. He identified two patterns of environmentally driven conflict: elite resource capture and ecological marginalisation, triggering migration to ecologically fragile areas. Importantly, these routes to conflict both involve elite and economic interests in their fulfilment (Homer-Dixon, 1994). Homer-Dixon cites the Moorish appropriation of newly fertile farmland in Mauritania, which resulted in skirmishes between the nation and Senegal, as evidence of violence stemming from elite resource capture. However, the problems that created this conflict lay beyond the events of resource capture that provided an immediate cause. Anne Guest (1995) reported that such origins can be found in the struggle to develop a Mauritanian national identity in the post-independence era, which resulted in the patronage of the Moorish majority and the political and economic exclusion of the African minority. Similar events occurred in Aceh, Indonesia, where a separatist struggle, although enveloped in a situation of resource capture, occurred in the context of national identity construction (Aspinall, 2007).

It is important not only to understand these causal frameworks but also to explore the different ways that the environment is perceived and interpreted by various people, communities and nations. Much of the scholarship on the topic has failed to explore important episodes of adaptation conflict – in which nature is transformed as a means to mitigate or adapt to future climatic change. However, what one individual perceives as a route to sustainable energy and development, another perceives as the ecological transformation of a river and the displacement of local communities. Important examples of this can be found in the struggles of indigenous peoples to protect their lands.

Such conflicts – primarily over forests and logging operations – are present across the world, from Brazil to Indonesia. For example, the unprecedented scale of logging operations in the Indonesian-controlled Kalimantan region of Borneo have led to intense conflicts between the native Dayak tribe and Indonesian forces (Klare, 2001). The subsequent Sampit conflict of 2001 resulted in the murder and decapitation of 100 Madurese migrants by the Dayak, before the Indonesian military restored control (BBC, 2001). However, although this conflict was triggered by state-initiated environmental degradation, the Dayak fought not for environmental reasons but for cultural and economic purposes, as the forest provided them with their livelihood and their societal base. An indigenous letter to the Indonesian authorities stated, 'Stop destroying the forest or we will be forced to protect it. The forest is our livelihood' (cited in Klare, 2001: 205). Indigenous action was not motivated by the intrinsic nature of the environment but by its cultural role – the destruction posing an existential threat.

The case of environmental migrants

The importance of competing variables, forming part of a greater causal web, is well illustrated in the case of ecological migrants (Homer-Dixon, 1994). It is stated that environmental refugees place strains on their host communities, undermining the ability of local governments to provide the necessary services (Salehyan, 2008) and resulting in threats posed to collective identity and social cohesion. Aning and Atta-Asamoah (2011) found that the migration from neighbouring states into Cote d'Ivoire was a major factor in the conflict that later engulfed the nation in a cycle of violence. However, the narrow assertion that these conflicts are the direct result of migration – itself exclusively ecologically induced – is misguided. Migration is the result of both push and pull factors and the environment is only one element in a greater causal framework.

The example provided by Homer-Dixon (1994) of environmental migration triggering violent conflict was that associated with Bengali migrants to the Assam region of India, which resulted in the 'Assam Agitation' – a popular movement against undocumented migrants. Although this movement was principally peaceful, cases of extreme violence did occur, including the massacre in 1983 of approximately 1,700 Bengalis in the village of Nellie (Homer-Dixon, 1994). However, mass migration from Bangladesh to neighbouring India was the result of a plethora of factors that pushed and pulled many towards emigration, including widespread impoverishment, ethnic divisions, patterns of private ownership, and institutional encouragement from many Indian politicians who were keen for more voters (Homer-Dixon, 1994; Lee, 2001). Therefore, such migration was not exclusively the result of ecological factors; it was also the consequence of socio-political issues and the failure of the Bangladeshi government to provide adequate livelihoods and opportunities to marginalised populations.

Furthermore, the 'Soccer War' of July 1969, fought between El Salvador and Honduras, is regularly cited as an example of the role that migration, caused by economic and ecological marginalisation, can have in the occurrence of international conflict (Dabelko et al., 2000). Although triggered by crowd violence at an international football match, the tensions surrounding this conflict were generated by a 1962 Honduran land reform act that involved the seizure of land from Salvadoran migrants and its redistribution to native Hondurans. However, it would be reckless to place exclusive causal importance for this conflict on environmental factors, which were not standalone but acted in tandem with social, economic and legal causes. The migratory patterns that caused this legislative response were the result not of strictly ecological factors but rather the inequality of land distribution within El Salvador itself (Kapuściński, 2007).

Furthermore, the resultant conflict was also predominantly motivated by socioeconomic factors. Durham (1979) found that the scarcity of land in Guatemala, frequently cited as a factor, had only limited causal importance, with the rapid expansion of commercial agriculture the cause of the ecological marginalisation of Guatemalans and Salvadorans alike. Political conflict between landowners and the *campesinos* created pressure on Honduran president General Lopez Arellano to protect the rights of landowners – resulting in pro-landowner land reform legislation that disproportionately affected the poor and displaced thousands of both Hondurans and migrant Salvadorans. It was this displacement of the diaspora in Honduras that provoked the strong reaction in El Salvador that ultimately led to the conflict. The Soccer War was not to do with ecology but with political and economic causality – the ecological marginalisation of *campesinos*, of both Honduran and Salvadoran nationality, the result of an economic contest between commercial landowner and peasant farmer.

The thesis, put forward by Myers and Kent (1995), that environmental refugees cause an upsurge in internal tensions resulting in violence in their host regions is flawed. The hypothesis was based on the assumption of a direct causal linkage, utilising examples of Ethiopia, Somalia, Nigeria, Bangladesh and Sudan, among others. The deduction that *Nation A* suffered from environmental degradation, experienced noticeable migratory patterns, and was subjected to violence ignores the wider contextual factors that must be addressed (Castles, 2002). Such an assertion fails to understand the cultural characterisation of migration in many regions. Examples of migration in the Sahel region as the result of displacement via desertification ignores the fact that such migration has been present for centuries as a strategy of community adaptation and the continuation of livelihoods (Black, 2001). Similarly, the context of the receiving nations provides an important barrier to potential violence. For example, the Asian tsunami of 2004 resulted in unprecedented levels of population displacement and scarcity but did not trigger conflict in Southeast Asia. This is due to the presence of accommodating frameworks in the receiving nations and areas, with social integration policies being particularly important. The existence of such systems has resulted in generalised violence against refugees becoming a rare event (Onoma, 2013).

Concluding remarks

The historical evidence fails to support any arguments of environmentally driven conflict and instead provides indications that the environment as an operating cause is placed within a causal framework where numerous factors intersect. Within the literature, a confused permutation of conflicts involving resources with an economic value and environmental degradation results in misguided conclusions regarding the environmental causality and nature of many conflicts. Such assumptions result in the neglect of additional variables that have created a framework in which conflict takes place. Social, political and economic factors are all important facets of the process that may eventually lead to conflict. It is these factors that have resulted in minerals in Angola and Sierra Leone leading to prolonged civil conflicts and unprecedented violence, and the same resources resulting in relative peace and growth in nations such as Botswana. The landmark academic examples of the role of such resources in the formation of conflicts neglect key variables in the occurrence of such struggles. World War Two was not caused by Adolf Hitler's pursuit of *lebensraum* alone. Numerous additional factors provided an underlying interplay that resulted in the outbreak of war in 1939. This chapter has sought to illustrate that such a relationship is particularly evident in the discourse of conflict surrounding environmental migrants and refugees.and policy do not explore these linkages further, the environmental conflict thesis may just become a self-fulfilling prophecy.

The consequence of this is not the total irrelevance of environmental factors in conflict but the etymological transition from environmentally *driven* to environmentally *induced* conflicts. Ecology will always possess a role in the causality of conflict and, with the spectre of further degradation, greater scarcity and the phenomena of climate change, this role can only increase. Although the societal effects of climate change in the future are fundamentally uncertain (Salehyan, 2008), it is important for academics and policymakers to understand that ecological factors are not the sole actor in the formation of patterns of conflict. Instead, they are part of a complex web of causality, coinciding with important social, political and economic factors that can result in both the presence and absence of violence. In order to truly understand this framework, simple determinism must be rejected and a more nuanced approach developed.

Lastly, a number of studies have pointed to the framing of climate change as a security issue as an important influence on events, with the perception by actors of *climate as security* potentially leading to more militarised responses (Salehyan 2008; Feitelson et al., 2012). With this in mind, it is important to be careful when discussing the links between the environment, resources and degradation, and conflict. Although there is an increased consciousness of environmental problems in many parts of the world, popular understanding of the complex relationships between climate change, violence and the labyrinth of other factors in the creation of conflict is still relatively inadequate. Future scholarship must transfer its focus away from conflicts over scarcity or abundance and towards issues of food and water security, livelihoods and development. It is these relationships between purely social causes of conflict and environmental factors that result in the upsurge of violence in a society responding to ecological degradation and stress. It is no longer possible to see environmental issues as neutral, detached from the social world – they must be understood as political problems with social drivers. If scholarship and policy do not explore these linkages further, the environmental conflict thesis may just become a self-fulfilling prophecy.

References

Aning, K. and Atta-Asamoah, A. (2011). *Demography, Environment and Conflict in West Africa*. Accra: Kofi Annan International Peacekeeping Training Centre.

Aspinall, E. (2007). The Construction of Grievance. *Journal of Conflict Resolution*, 51(6), 950-972.

Bannon, I. and Collier, P. (eds) (2003). *Natural Resources and Violent Conflict: Options and Actions*. Washington, DC: The World Bank.

Benjaminsen, T., Alinon, K., Buihaug, H. and Buseth, J. T. (2012). Does Climate Change Drive Land-use Conflicts in the Sahel? *Journal of Peace Research*, 49(1), 97-111.

Barnett, J. (2000). Destabilising the Environment-Conflict Thesis. *Review of International Studies*, 26(2), 271-288.

Barnett, J. and Adger, A. (2007). Climate Change, Human Security and Violent Conflict. *Political Geography*, 26(6), 629-655.

BBC (2001, February 27). *Horrors of Borneo Massacre Emerge*. Retrieved from http://news.bbc.co.uk/1/hi/world/asia-pacific/1191865.stm.

Black, R. (2001). Environmental Refugees: Myth or Reality. *Journal of Humanitarian Assistance*. Working Paper No. 34. Geneva: UNHCR. Retrieved from http://www.unhcr.org/3ae6a0d00.pdf

Castles, S. (2002). *Environmental Change and Forced Migration: Making Sense of the Debate*. UNHCR: Issues in Refugee Research, Working Paper No. 70. Geneva: UNHCR. Retrieved from http://www3.hants.gov.uk/forced_migration.pdf

Collier, P. and Hoeffler, A. (2004) Greed and Grievance in Civil War. *Oxford Economic Papers*, 56(4), 563-595.

De Koning, R. (2008). *Resource-Conflict Links in Sierra Leone land the Democratic Republic of Congo*. Stockholm: Stockholm International Peace Research Institute.

Dabelko, G. D. and Dabelko, D. D. (1993). *Environmental Security: Issues of Conflict and Redefinition*. Harrison Programme on the Future Global Agenda, Occasional Paper No.1. College Park, MD: University of Maryland.

Dabelko, G., Long, S. and Matthew, R. (2000). *State-of-the-Art Review of Environment, Security and Development Cooperation*. OECD Working Paper. Retrieved from https://www.iisd.org/pdf/2002/envsec_oecd_review.pdf

Dalby, S. (2002). *Environmental Security*. Minneapolis: University of Minnesota Press.

Deudney, D. (1990). The Case Against Linking Environmental Degradation and National Security. *Millennium: Journal of International Studies*, 19(3), 461-476.

Dokken, K. and Graeger, N. (1995). *The Concept of Environmental Security – Political Slogan or Analytical Tool?* PRIO Report, No. 2. Oslo: International Peace Research Institute.

Durham, W. H. (1979). *Scarcity and Survival in Central America: Ecological Origins of the Soccer War*. Stanford: Stanford University Press.

Elliott, L. (1998). *The Global Politics of the Environment*. London: Macmillan.

Evans, A. (2010). *Resource Scarcity, Climate Change and the Risk of Violence Conflict*. Washington, DC: World Bank. Retrieved from http://documents.worldbank.org/curated/en/2010/09/14296670/resource-scarcity-climate-change-risk-violent-conflict

Feitelson, E., Tamimi, A. and Rosenthal, G. (2012). Climate Change and Security in the Israeli-Palestinian Context. *Journal of Peace Research*, 49(1), 241-257.

German Advisory Council on Global Change (WBGU) (2007). *Climate Change as a Security Risk*. Berlin: Earthscan. Retrieved from http://www.wbgu.de/fileadmin/templates/dateien/veroeffentlichungen/hauptgutachten/jg2007/wbgu_jg2007_kurz_engl.pdf

Gleditsch, N. P. (1998). Armed Conflict and the Environment: A Critique of the Literature. *Journal of Peace Research*, 35(3), 381-400.

Guest, A. (1995). Conflict and Cooperation in a Context of Change: A Case Study of the Senegal River Basin. In: J. MacMillan and A. Linklater (eds). *Boundaries in Question: New Directions in International Relations*. London: Pinter.

Haas, P. M. (2002). Constructing Environmental Scarcity. *Global Environmental Politics*, 2(1), 1-11.

Hauge, W. and Ellingsen, R. (1998). Beyond Environmental Scarcity: Causal Pathways to Conflict. *Journal of Peace Research*, 35(3), 227-243.

Homer-Dixon, T. F. (1994). Environmental Scarcities and Violent Conflicts: Evidence from Cases. *International Security*, 19(1), 5-40.

Hsiang, S. M., Burke, M. and Miguel, E. (2013). Quantifying the Influence of Climate on Human Conflict. *Science*, 341 (6151).

Kapuściński, R. (2007). The Soccer War. London: Granta Books.

Klare, M. T. (2001). *Resource Wars: The New Landscape of Global Conflict*. New York: Henry Holt & Company.

Le Billon, P. (2001). The Political Ecology of War: Natural Resources and Armed Conflicts. *Political Geography*, 20(5), 561-583.

Le Billon, P. (2012). *Wars of Plunder: Conflict, Profits and the Politics of Resources*. London: Hurst & Co.

Lee, S.-W. (2001). *Environment Matters: Conflict, Refugees and International Relations*. Seoul and Tokyo: World Human Development Institute Press.

Libiszewski, S. (1992). *What is an Environmental Conflict?* ENCOP Occasional Paper No. 1. Berne: Swiss Peace Foundation.

Lipschutz R. and Holdren, J. (1990). Crossing Borders: Resource Flows, the Global Environment and International Stability. *Bulletin of Peace Proposals*, 21(2), 121-133.

Lujala, P. (2010). The Spoils of Nature: Armed Civil Conflict and Rebel Access to Natural Resources. *Journal of Peace Research*, 47(1), 15-28.

Meyer, L., Lohr, C., Gronenborn, D. and Alt, K. W. (2015). The Massacre Mass Grave of Schöneck-Kilianstädten Reveals New Insights into Collective Violence in Early Neolithic Central Europe. *Proceedings of the National Academy of Sciences*, 112(36), 11217-11222.

Moorehead, R. (1992). Land Tenure and Environmental Conflict: The Case of the Inland Niger Delta. In: J. Käkönen (ed.). *Perspectives on Environmental Conflict and International Relations* (pp. 96-115). London: Pinter.

Myers, N. and Kent, J. (1995). *Environmental Exodus: An Emergent Crisis in the Global Arena*. Washington, DC: The Climate Institute. Retrieved from http://www.climate.org/PDF/Environmental%20Exodus.pdf

Onoma, A. K. (2013). *Anti-Refugee Violence and African Politics*. Cambridge: Cambridge University Press.

Selby, J. (2014). Positivist Climate Conflict Research: A Critique. *Geopolitics*, 19(4), 829-856.

Salehyan, I. (2008). From Climate Change to Conflict? No Consensus Yet. *Journal of Peace Research*, 45(3), 315-326.

Smith, D. and Vivenkananda, J. (2007). *A Climate of Conflict: The Links between Climate Change, Peace and War*. London: International Alert.

UNEP (2009). *From Conflict to Peacebuilding: The Role of Natural Resources and the Environment*. Nairobi: United Nations Environment Programme.

Verhoeven, H. (2011). Climate Change, Conflict and Development in Sudan: Global Neo-Malthusian Narratives and Local Power Struggles. *Development and Change*, 42(3), 679–707.

Westing, A. J. (ed.) (1986). *Global Resources and International Conflict: Environmental Factors in Strategic Policy and Action*. Oxford: Oxford University Press.

Wolf, A. T. (1998). Conflict and Cooperation along International Waterways. *Water Policy*, 1(2), 251-265.

8

Actors other than States: The Role of Civil Society and NGOs as Drivers of Change

EMILIE DUPUITS
UNIVERSITY OF GENEVA, SWITZERLAND

On 21 September 2014, around 300,000 people in New York City participated in the largest climate change demonstration in history, calling for climate justice and action (Foderaro, 2014). This event reflects the increasing mobilisation of civil society and non-governmental organisations (NGOs) in global environmental processes. It also raises questions about democratic decision-making and participation processes facing the deficit of governments' action, especially in climate change international negotiations. This involvement suggests the following question: To what extent are civil society and NGOs drivers of change in global environmental governance?

This chapter does not pretend to provide a detailed analysis of all the aspects related to civil society and NGOs' international involvement in global environmental governance but rather to give an overview on some major issues for future research on the topic, in a transdisciplinary perspective. In addition, the author intends to provide some responses to the interrogation, through four main sections. First, it is necessary to understand the context of global environmental governance in which civil society actors emerge. Indeed, global environmental governance can be defined as a multi-level (i.e. global nature of environmental problems and local impacts), multi-actor (i.e. states, experts, environmental NGOs, and individuals) and multi-sector (i.e. energy, water and trade) approach that represents both an opportunity and a constraint for civil society actors. Second, the concept of transnational civil society is carefully defined to grasp the particularities of the multiple actors constituting this category. In fact, rather than a unified category, transnational civil society is composed of a plurality of actors, ranging from environmental NGOs, epistemic communities and social movements to civil society organisations. The latter receives special attention jointly with new processes of change in global environmental governance. Third, a literature review presents authors from different theoretical traditions who have studied the involvement of civil society actors in international environmental governance. These authors have provided different interpretations of actors' roles – from marginal to central – and impacts – from external impacts on states to internal impacts on own actors – around key issues such as legitimacy and democracy.

Finally, two main processes of change related to civil society involvement in global environmental governance are analysed: internationalisation and autonomisation. On one hand, the international involvement of civil society actors questions the strategies employed to reframe existing global norms and rights, the instrumentalisation of overlaps or missing links between international regimes and sectors, and the legitimation of international representativeness. On the other hand, the civil society's internationalisation produces effects back on their level of autonomy, through processes of professionalisation and expertise-building.

Analysing the actors involved in multi-scalar environmental governance implicates a dialogue between various theoretical disciplines, including international relations, political science, sociology and geography. This chapter aims principally to centre the analysis not on institutions and formal processes of environmental regulation but on actors, their interactions and scalars politics, influencing the construction of environmental issues and their modes of resolution.

The rising power of civil society in global environmental governance

In the 1970s, environmental issues began to be included in global governance architectures, especially with the first United Nations (UN) Conference on the Human Environment in Stockholm in 1972. These environmental issues are of different natures, ranging from global commons, such as ozone layer, climate change or biodiversity conservation, to local commons, such as water depletion or deforestation (Young et al., 2006; Ostrom, 2010). In the field of international relations, the environment has been progressively analysed as an important driver of change in multilateral negotiations through the participation of multiple actors, such as nation-states, international organisations and NGOs (O'Neill, 2009).

However, the difficulties in setting up a global agreement on climate change and the limits to establishing global regimes on natural resources regulation – as in the cases of water and forests – demonstrate the high fragmentation of global environmental governance that currently exists (Gupta and Pahl-Wostl, 2013; Biermann et al., 2009; Giessen, 2013). This represents, on one hand, an opportunity for civil society and NGOs to enter global arenas relatively opened to non-state actors' participation. The World Summit on Sustainable Development of Johannesburg in 2002 provided insight into the increased participation of actors from civil society through innovative processes of multi-stakeholder deliberation and public-private partnerships (Bernstein, 2012). On the other hand, this high fragmentation also represents a constraint, as the efforts to dominate norm-building processes by the multiplication of civil society actors involved in international processes can generate power relations and competition (Andonova and Mitchell, 2010). Disagreements among these actors upsets the appropriate scale at which to govern natural resources and causes diverging representations on the nature of resources (from public to economic goods, or territorial to universal rights). Some authors prefer to discuss multi-scalar governance in terms of the shift from a hierarchical international system towards a horizontal network system (Cash et al., 2006). Network governance is characterised by the multiplication of interrelations between actors at different scales (Diani and McAdam, 2003; Bulkeley, 2005). While the concept of multi-scales governance acknowledges the key importance of nation-states in norm-building and decision-making processes, it also attributes an important role to non-state actors like NGOs, expert networks and CSOs. Andonova and Mitchell (2010: 256) differentiate two main processes of rescaled governance: 'global governance has been rescaled away from the nation-state in multiple directions: vertically down toward provincial and municipal governments, vertically up toward supranational regimes, and horizontally across regional and sectoral organisations and networks'.

Moreover, global norms and paradigms are the object of increasing transnational protests, mainly directed against the lack of civil society organisations' (CSOs) inclusion in decision-making processes (Conca, 2005; Visseren-Hamakers et al., 2012). Indeed, CSOs are often represented in global arenas through intermediaries such as international NGOs (McMichael, 2004; Vielajus, 2009; Siméant, 2010). The implementation of 'commodity consensus' on natural resources by international technical experts is also a major point of contestation from CSOs (Svampa, 2015). As an example, Conca (2005) stresses the paradox of global water governance, characterised by intents from experts' networks and international NGOs to create institutionalised norms and blueprints, in parallel with the rising contentions from less visible civil society actors.

Some authors prefer to discuss multi-scalar governance in terms of the shift from a hierarchical international system towards a horizontal network system (Cash et al., 2006). Network governance is characterised by the multiplication of interrelations between actors at different scales (Diani and McAdam, 2003; Bulkeley, 2005). While the concept of multi-scales governance acknowledges the key importance of nation-states in norm-building and decision-making processes, it also attributes an important role to non-state actors like NGOs, expert networks and CSOs. Andonova and Mitchell (2010: 256) differentiate two main processes of rescaled governance: 'global governance has been rescaled away from the nation-state in multiple directions: vertically down toward provincial and municipal governments, vertically up toward supranational regimes, and horizontally across regional and sectoral organisations and networks'.

In a vertical perspective, civil society actors can mobilise the international scale to build environmental problems as global (i.e. deforestation and environmental migrations), or conversely, to increase their international visibility so as to respond to some local environmental issues (i.e. extractive conflicts and drinking water access). In a horizontal perspective, marginalised civil society actors can gain more power through the creation of transnational networks or multi-stakeholders partnerships (i.e. transnational indigenous and peasant networks). The concept of multi-scales governance is useful for revealing how actors mobilise simultaneously in different jurisdictional (i.e. national, regional and local), territorial (i.e. village and hydrographic basin) and sectorial (i.e. energy and trade) scales (Compagnon, 2010). It is also useful for understanding the strategies employed to influence this multi-layered decision-making structure. As developed further in this chapter, the analysis of norms and discourses is of particular importance for understanding power relations happening within the transnational civil society. Particular attention is given to actors playing the role of intermediaries or brokers between scales. In the next part, different types of actors within transnational civil society are shown to highlight the pluralism of actors and definitions.

Searching for a transnational civil society

The transnational approach inside international relations theory highlights the dynamics of actors intervening not only around nation-states but at multiple scales (Keohane and Nye, 1972; Rosenau, 2002). Transnational relations can be defined as 'regular interactions across national boundaries where at least one actor is a non-state agent or does not operate on behalf of a national government or an intergovernmental organisation' (Risse-Kappen, 1995: 3). The transnational perspective provides a more dynamic view on civil society actors, considered both as agents and subjects of change

(Khagram et al., 2002). It also breaks with the idea that decisions taken at the international scale would apply automatically at inferior scales with a cascade effect. Instead of a truly global or local scale, it is more suitable to speak of a continuum of interactions with an intrinsic dynamic structure.

Various authors have provided definitions of transnational, global and international civil society, with the objective of building a unified category of analysis. Some authors analysed civil society and NGOs as agents of coordination and pacified collaboration at the international scale (Kaldor, 2003; Keane, 2003). Others point out the difficulty of delimiting actors' boundaries in one single category, illustrated by the question of whether or not to include private businesses as civil society actors (Gemmill and Bamidele-Izu, 2002). In the following paragraphs, the key characteristics and differences of four main types of actors are described. These actors do not comprise an exhaustive list but provide an interesting overview of civil society involvement in global environmental governance.

With the multiplication of global environmental conferences in the 1970s and treaty negotiations in the 1990s, various NGOs increased their international involvement. The first major international NGOs (INGOs) emerged in the field of biodiversity and forest conservation, such as the World Wildlife Fund (WWF), Greenpeace and the International Union for the Conservation of Nature (IUCN). These INGOs started to defend the environmental value of forests at international level, justifying the creation of protected areas. In turn, these actors became part of a wider global context focusing on awareness promotion around environmental damages, such as large deforestation in the Amazon and species massive extinction (Epstein, 2008). They have also played the role of the primary intermediaries and representatives of marginalised actors, like indigenous peoples and local communities (Dumoulin, 2003; Aubertin, 2005).

Within international relations, another interesting category of actors are the transnational advocacy networks (TANs). A TAN refers to 'those actors working internationally on an issue, who are bound together by shared values, a common discourse, and dense exchanges of information and services' (Keck and Sikkink, 1999: 89). This type of network plays an important role in the regulation of globalisation, seeking primarily to influence states and international organisations. In this sense, INGOs have participated in the process of global norms redefinition, highlighting their intermediary role between local actors and their global claims. Some examples of TANs emerging in the environmental field are linked to claims of global environmental justice, such as in the cases of the Yasuni ITT campaign to keep oil in the Ecuadorian Amazon (Martin, 2011), or the anti-Narmada dam movement in India (Conca, 2005).

Third, other groups of actors are emerging to fill the gap of scientific uncertainty and complexity related to global environmental problems. These epistemic communities are defined as 'a network of professionals with recognised expertise and competence in a particular domain and an authoritative claim to policy-relevant knowledge within that domain or issue-area' (Haas, 1992: 3). Technical experts are characterised by their professionalisation and authority in one domain, scientific knowledge and neutrality (Conca, 2005). An emblematic example of an epistemic community is the Intergovernmental Panel on Climate Change (IPCC), created in 1988 to respond to the high uncertainty and complexity of climate change phenomenon and its consequences.

Finally, actors from civil society organisations (CSO) are progressively integrated into global environmental processes through the creation of transnational grassroots networks (Smith and Guarnizo, 1998; Escobar, 2008). The particularity of these networks lies in their self-management and membership, as they are exclusively composed of grassroots organisations – defined as 'those who are most severely affected in terms of the material condition of their daily lives' (Batliwala, 2002: 396) – both providers and recipients of collective service, and therefore directly concerned by the issue they are defending. This concept echoes the idea of cosmopolitan localism (McMichael, 2004), referring to the active role local communities play to regain ownership of global issues that affects them directly. This can happen through the increased awareness of shared interests and values with other local actors previously isolated from each other (Caouette, 2010). An example of a transnational grassroots network is the Coordinator of the Indigenous Organisations of the Amazon Basin (COICA), created in 1984 to defend territorial rights and local autonomy.

Looking inside civil society: legitimacy and democracy

The multiplication of civil society actors and their interactions in global environmental governance raises questions about legitimacy and representativeness between local and international scales. Various authors have analysed legitimacy as the lack of institutions' authority in the field of environmental governance. They have argued for the creation of a World Environmental Organisation (Biermann and Bauer, 2005). In a critical sociological perspective, Bernstein (2012) analyses the multiplication of potentially competitive initiatives to fill the lack of authority and gain power in global environmental governance. He defines political legitimacy as 'acceptance and justification of shared rule by a community' (p. 148), emphasising the role of perceptions around the most legitimate actors and scales of action.

Actors' pluralism, and the differences in terms of power, invites us to deconstruct the unified category of transnational civil society. Indeed, most research has focused on the interactions between NGOs and governments, or between TANs and international organisations. However, internal interactions, strategies and conflicts among civil society actors and networks also matter in an analysis of changes to global environmental governance. Various studies point to the limits of ideal democratic character attributed to civil society actors in global environmental governance (Bäckstrand, 2012; Bernauer and Betzold, 2012). An interesting empirical example to demonstrate the existence of legitimacy issues among civil society actors is the demand from indigenous people networks of more autonomy and self-representation in relation to conservationist NGOs, with the aim of defending a more integrated vision of forest management (Aubertin, 2005).

Dumoulin and Pepis-Lehalleur (2012) define the network not as a structure but as a social object on behalf of instrumentalisations and representations. An interesting example is the recognition of the norm of food sovereignty from CSOs and its multiple interpretations. When indigenous people networks employ arguments about conservation; transnational peasant movements – such as Via Campesina – are more linked to sustainable exploitation and innovation perspective (Brenni, 2015). Regarding the category of epistemic communities, Haas (2015) recently acknowledged the need to redefine the concept so as to include the analysis of internal issues in the construction of a shared scientific knowledge.

Both international relations (IR) and social movement theory have focused on the analysis of transnational civil society's external impacts on states and decision-making processes (Tarrow and McAdam, 2004; Chabanet and Giugni, 2010). From a sociological perspective, transnational collective action has been defined as 'the coordinated international campaigns on the part of networks of activists against international actors, other states, or international institutions' (Della Porta and Tarrow, 2005: 7). For example, Smith (2008) provided an analysis of the democratic globalisation network, highlighting its failure to challenge the dominant neoliberal globalisation paradigm because of its lack of coherence and articulation. IR studies have also attributed a reduced role to civil society actors in the norm life-cycle model (Finnemore and Sikkink, 1998). In this sense, civil society actors participate in the early stages of norm emergence, through advocacy strategies, but not in the diffusion and internalisation phases.

Most of these approaches emphasise the contentious or reactive character of transnational social movements (Tarrow and McAdam, 2004; Della Porta and Tarrow, 2005). The focus of their action remains unconnected to the object of study, measuring effectiveness in terms of impacts on states and other international actors (Dufour and Goyer, 2009). To respond to these limits, other authors ask for the study of more sustainable networks implementing strategies other than protest (Vielajus, 2009; Siméant, 2010; Caouette, 2010). Saunders (2013) considers environmental networks as the outcome variable to be explained and not the factor to explain political impacts and changes. The analysis of transnational grassroots networks requires a deeper understanding of the diversity among – and within – members at the local scale and the degree of autonomy between scales of representativeness. The next section focuses particularly on these particular actors and issues.

New spaces of internationalisation and autonomisation

This part aims to analyse civil society's international involvement as a driver of change regarding global environmental politics. Some authors talk about the geographical turn in the study of social movements, focusing on the links between scale politics and transnational collective action. The concept of scale has been defined in the field of critical geography as the interactional process underlying power relations between actors (Swyngedouw, 1997; Masson, 2009). On one hand, participation of civil society actors at the international level questions the strategies employed to reframe existing global norms and rights, the instrumentalisation of overlaps or missing links between international regimes and sectors and the legitimation of an international representativeness (Cash et al., 2006). On the other hand, civil society's internationalisation produces effects back on their level of autonomy, through processes of professionalisation and expertise-building (Staggenborg, 2010).

Siméant (2010) defines three main axes of social movements' internationalisation: global framing, new repertories of strategies, and new forms of international representativeness. The first one is linked to framing strategies, defined as the 'strategic efforts by groups of people to fashion shared understandings of the world and of themselves that legitimate and motivate collective action' (Khagram et al., 2002: 12). Siméant (2009), in a general comment, and Martin (2011), in relation to Yasuni ITT, raise the interesting point that if civil society actors often rely on existing global norms, such as universal human rights, to defend their cause, they are also contributing to reframing it through alternative representations. An example is the reframing of human right to water into community-based right to water, to acknowledge the crucial role of communities in providing drinking water in rural areas (Bakker, 2007; Dupuits, 2014). Alternatively, local norms can be reframed as global, for example to build a common identity or gain more influence in higher decision-making arenas. An example is the fight of transnational indigenous and forest communities' networks for recognition of territorial rights, challenging the dominant paradigm of market-based environmental norms (Dupuits, 2015). Reframing strategies are particularly important in a context where 'discourses of expertise that are setting the rules for global transactions, even in the progressive parts of the international system, have left ordinary people outside and behind' (Appadurai, 2000: 2).

A second implication of internationalisation is the mobilisation of innovative strategies beyond protest. One major action strategy used by transnational civil society networks is their integration into fragmented international environmental regimes and sectors. Fragmentation refers to the notion of regime complex defined as 'a network of three or more international regimes that relate to a common subject matter; exhibit overlapping membership; and generate substantive, normative, or operative interactions recognised as potentially problematic whether or not they are managed effectively' (Orsini et al., 2013). Orsini (2013) developed the concept of 'multi-forum shopping' to define the capacity of civil society actors to shift participation and advocacy from one arena to another according to their receptiveness and to serve particular interests. On one hand, actors can contribute to integrating previously disconnected regimes and arenas. On the other hand, actors can intend to exacerbate existing overlaps between regimes and sectors to strengthen their claims. One illustrative example is the influence of transnational indigenous networks in identifying overlaps between the climate regime and the biodiversity regime (prioritisation of carbon over biodiverse forests, increase of local social inequalities) (Harrison and Paoli, 2012) to serve their interest of regaining control over climate global funds and territorial rights.

Global environmental arenas are increasingly challenged by new forms of representativeness, emerging mainly from the local level and more direct forms of collective action. However, participation of grassroots organisations at the international level does not resolve the legitimacy gap mentioned in the previous section. On the contrary, some transnational grassroots networks tend to reproduce practices of depoliticisation. Wilson and Swyngedouw (2014: 56) define this process as a 'global, borderless regime where rules are formulated by panels of technocrats and framed in neutralised terms of standards settings and harmonization.' Depoliticisation is used to justify the possibility of representing really diverse local actors by unifying them in a common category and claiming new expertise.

Finally, the rising inclusion of CSOs in international decision-making arenas implies new spaces of autonomisation, including processes of professionalisation and alternative forms of expertise. Staggenborg (2010) differentiates 'classical movement organisations, which rely on the mass mobilization of "beneficiary" constituents as active participants, [and] "professional" social movement organisations (SMOs) relying primarily on paid leaders and "conscience" constituents who contribute money and are paper members rather than active participants' (599). Siméant (2010) also talks about a rising trend towards the NGO-isation of grassroots movements. An illustrative example is the increasing inclusion of civil society actors as key stakeholders in global climate decision-making arenas. Indeed, under the international climate change regime was launched in 2008 the UN-REDD (Reducing Emissions from Deforestation and forest Degradation) programme. UN-REDD, as an emerging powerful technical expert, aims to fight deforestation by creating a financial value for carbon stored in forests through market mechanisms (McDermott et al., 2012). The rising demand for articulated, formalised and representative CSOs in UN-REDD decision-making processes is an example of new dynamics of professionalisation (Wallbott, 2014). As a product of this professionalisation dynamic, a new type of local and grassroots expertise is emerging in global environmental arenas, crossing both expert and militant logics (Foyer, 2012). Grassroots expertise refers to 'a wide range of practical skills and accumulated experience, though without any formal qualifications' (Jenkins, 2009: 880). Studying changes to global environmental governance through the lenses of transnational civil society networks reveals a variety of innovations that need further research.

Conclusion

This chapter has shown the importance of considering the context and plurality of actors and theoretical approaches to analysing civil society and NGOs' international involvement in global environmental governance. Indeed, this involvement occurs in the context of complex multi-scales environmental governance in terms of decision-making levels, actors and sectors. The short presentation of actors constituting transnational civil society demonstrates the great plurality of actors and the impossibility of defining a unified category. This raises theoretical concerns related to issues of legitimacy, democracy and internal conflicts. Finally, the last section aimed to analyse some modalities of change in global environmental governance brought by civil society actors, beyond the analysis of effectiveness and external democracy. Grassroots organisations have been given a particular focus in order not to reproduce the over-analysis on NGOs and other international powerful actors.

Five major changes have been identified: the reframing of global norms with alternative interpretations and concrete modalities of implementation, the connection of fragmented international environmental regimes as a tool of integration or pressure, the definition of new forms of international grassroots representativeness beyond protest strategies and linked to processes of depoliticisation, the professionalisation of civil society through the inclusion to complex international decision-making arenas, and the construction of new grassroots expertise. These changes demonstrate the existence of a twofold impact: of civil society's international involvement on global environmental governance, and of internationalisation processes on civil society forms of autonomy and expertise.

References

Appadurai, A. (2000). Grassroots Globalization and the Research Imagination. *Public Culture*, 12(1), 1-19.

Andonova, L. B. and Mitchell, R. B. (2010). The Rescaling of Global Environmental Politics. *The Annual Review on Environment and Resources*, 35, 255-282.

Aubertin, C. (ed.) (2005). *Représenter la nature? ONG et biodiversité*. Paris: IRD Editions.

Bäckstrand, K. (2012). Democracy and global environmental politics. In P. Dauvergne (ed.). *Handbook of Global Environmental Politics*. 2nd Edition (pp. 507-519). Cheltenham: Edward Elgar Publishing.

Bakker, K. (2007). The 'Commons' versus the 'Commodity': Alter-Globalization, Anti-Privatization, and the Human Right to Water in the Global South. *Antipode*, 39(3), 430-455.

Batliwala, S. (2002). Grassroots Movements as Transnational Actors: Implications for Global Civil Society. *Voluntas: International Journal of Voluntary and Nonprofit Organizations*, 13(4), 393-409.

Bernauer, T. and Betzold, C. (2012). Civil Society in Global Environmental Governance. *The Journal of Environment and Development,* 21(1), 62-66.

Bernstein, S. (2012). Legitimacy Problems and Responses in Global Environmental Governance. In P. Dauvergne (Ed.). *Handbook of Global Environmental Politics*. 2nd Edition (pp. 147-162). Cheltenham: Edward Elgar Publishing.

Biermann, F., Pattberg, P., van Asselt, H. and Zelli, F. (2009). The Fragmentation of Global Governance Architectures: A Framework for Analysis. *Global Environmental Politics*, 9(4), 14-40.

Biermann, F. and Bauer, S. (eds) (2005). *A World Environment Organization. Solution or Threat for Effective International Environmental Governance?* Aldershot: Ashgate.

Brenni, C. (2015). Where Are the Local Communities? Food Sovereignty Discourse on International Agrobiodiversity Conservation Strategy. In A. Trauger (ed.). *Food Sovereignty in International Context: Discourse, Politics and Practice of Space* (pp. 15-34). Abingdon: Routledge.

Bulkeley, H. (2005). Reconfiguring Environmental Governance: Towards a Politics of Scales and Networks. *Political Geography*, 24(8), 875-902.

Caouette, D. (2010). Globalization and Alterglobalization: Global Dialectics and New Contours of Political Analysis? *Kasarinlan: Philippine Journal of Third World Studies*, 25(1-2), 49-66.

Cash, D. W., Adger, W. N., Berkes, F., Garden, P., Lebel, L., Olsson, P., Pritchard, L. and Young, O. (2006). Scale and Cross-Scale Dynamics: Governance and Information in a Multilevel World. *Ecology and Society*, 11(2), art. 8. Retrieved from http://www.ecologyandsociety.org/vol11/iss2/art8/

Chabanet, D. and Giugni, M. (2010). Les conséquences des mouvements sociaux. In E. Agrikoliansky, O. Fillieule, and I. Sommier (eds). *Penser les mouvements sociaux. Conflits sociaux et contestations dans les sociétés contemporaines* (pp. 145-161). Paris: La Découverte.

Compagnon, D. (2010). Les défis politiques du changement climatique: de l'approche des régimes internationaux à la gouvernance transcalaire globale. In C. Cournil and C. Colard-Fabregoule (Eds). *Changements climatiques et défis du droit* (pp. 31-49). Brussels: Bruylant.

Conca, K. (2005). *Governing Water. Contentious Transnational Politics and Global Institution Building*. Cambridge, MA: The MIT Press.

Della Porta, D. and Tarrow, S. (eds) (2005) *Transnational Protest and Global Activism: People, Passions, and Power*. Lanham, MD: Rowman and Littlefield Publishers.

Diani. M. and McAdam, D. (eds) (2003) *Social Movements and Networks: Relational Approaches to Collective Action*. Oxford: Oxford University Press.

Dufour, P. and Goyer, R. (2009). Analyse de la transnationalisation de l'action collective: proposition pour une géographie des solidarités transnationales. *Sociologie et sociétés*, 41(2), 111-134.

Dumoulin, D. (2003). Les savoirs locaux dans le filet des réseaux transnationaux d'ONG: perspectives mexicaines. *Revue internationale des sciences sociales*, 178, 655-666.

Dumoulin, D. and Pepis-Lehalleur, M. (eds) (2012). *Agir-en-réseau. Modèle d'action ou catégorie d'analyse?* Rennes: Presses Universitaires de Rennes.

Dupuits, E. (2014). Construire une norme transnationale en réseau: gestion communautaire de l'eau et associativité en Amérique latine. *Revue Interdisciplinaire de Travaux sur les Amériques*, 7. Retrieved from http://www.revue-rita.com/dossier7/construire-une-norme-transnationale-en-reseau-gestion-communautaire-de-l-eau-et-associativite-en-amerique-latine.html

Dupuits, E. (2015). Transnational Self-help Networks and Community Forestry: A Theoretical Framework. *Forest Policy and Economics*, 58, 5-11.

Epstein, C. (2008). *The Power of Words in International Relations: Birth of an Anti-whaling Discourse*. Cambridge, MA: MIT Press.

Escobar, A. (2008). *Territories of Difference: Place, Movements, Life, Redes*. Durham, NC: Duke University Press.

Finnemore, M. and Sikkink, K. (1998). International Norm Dynamics and Political Change. *International Organization*, 52(4), 887-917.

Foderaro, L. W. (2014, September 21). Taking a Call for Climate Change to the Streets. *The New York Times*. Retrieved 21 October 2015, from http://www.nytimes.com/2014/09/22/nyregion/new-york-city-climate-change-march.html?_r=1

Foyer, J. (2012). Le réseau global des experts-militants de la biodiversité au cœur des controverses sociotechniques. *Hermès, La Revue*, 64, 155-163.

Gemmill, B. and Bamidele-Izu, A. (2002). The Role of NGOs and Civil Society in Global Environmental Governance. In D. Esty, and M. Ivanova (Eds). *Global Environmental Governance: Options and Opportunities* (pp. 77-100). Princeton, NJ: Yale School of Forestry and Environmental Studies.

Giessen, L. (2013). Reviewing the Main Characteristics of the International Forest Regime Complex and Partial Explanations for its Fragmentation. *International Forestry Review*, 15(1), 60-70.

Gupta, J. and Pahl-Wostl, C. (2013). Global Water Governance in the Context of Global and Multilevel Governance: Its Need, Form, and Challenges. *Ecology and Society*, 18(4), art. 53. Retrieved from http://www. ecologyandsociety.org/vol18/iss4/art53/

Haas, P. M. (1992). Introduction: Epistemic Communities and International Policy Coordination. *International Organization*, 46(1), 1-35.

Haas, P. M. (2015) *Epistemic Communities, Constructivism and International Environmental Politics*. Abingdon: Routledge.

Harrison, M. E. and Paoli, G. D. (2012). Managing the Risk of Biodiversity Leakage from Prioritising REDD+ in the Most Carbon-rich Forests: The Case Study of Peat-swamp Forests in Kalimantan, Indonesia. *Tropical Conservation Science*, 5(4), 426-433.

Jenkins, K. (2009). Exploring Hierarchies of Knowledge in Peru: Scaling Urban Grassroots Women Health Promoters' Expertise. *Environment and Planning A*, 41(4), 879-895.

Kaldor, M. (2003). The Idea of Global Civil Society. *International Affairs*, 79(3), 583-593.

Keane, J. (2003). *Global Civil Society?* Cambridge: Cambridge University Press.

Keck, M. E. and Sikkink, K. (1999). Transnational Advocacy Networks in International and Regional Politics. *International Social Science Journal*, 51(159), 89-101.

Keohane, R. O. and Nye Jr., J. (1972). *Transnational Relations and World Politics*. New Haven, CT: Harvard University Press.

Khagram, S., Riker, J. V. and Sikkink, K. (2002). *Restructuring World Politics: Transnational Social Movements, Networks and Norms*. Minneapolis: University of Minnesota Press.

Martin, P. (2011). Global Governance from the Amazon: Leaving Oil Underground in Yasuni National Park, Ecuador. *Global Environmental Governance*, 11(4), 22-42.

Masson, D. (2009). Politique(s) des échelles et transnationalisation: perspectives géographiques. *Politique et Sociétés*, 28(1), 113-133.

McDermott, M., Mahanty, S. and Schreckenberg, K. (2012). Examining equity: A Multidimensional Framework for Assessing Equity in Payments for Ecosystem Services. *Environmental Science and Policy*, 33, 416-427.

McMichael, P. (2004). *Development and Social Change. A Global Perspective*. London: Sage.

O'Neill, K. (2009). *The Environment and International Relations*. Cambridge: Cambridge University Press.

Orsini, A. (2013). Multi-forum Non-state Actors: Navigating the Regime Complexes for Forestry and Genetic Resources. *Global Environmental Politics*, 13(3), 34-55.

Orsini, A., Morin, J-F. and Young, O. R. (2013). Regime Complexes: A Buzz, a Boom or a Boost for Global Governance? *Global Governance*, 19, 27-39.

Ostrom, E. (2010). Polycentric Systems for Coping with Collective Action and Global Environmental Change. *Global Environmental Change*, 20(4), 550-557.

Risse-Kappen, T. (ed.) (1995). *Bringing Transnational Relations Back-in: Non-State Actors, Domestic Structures and International Institutions*. Cambridge: Cambridge University Press.

Rosenau, J. (2002). Governance in a New Global Order. In D. Held, and A. McGrew (eds). *Governing Globalization: Power, Authority and Global Governance* (pp. 70-86). Cambridge: Polity.

Saunders, C. (2013). *Environmental Networks and Social Movement Theory*. London: Bloomsbury Academic.

Siméant, J. (2009). Transnationalisation/internationalisation. In O. Fillieule, L. Mathieu, and C. Péchu (eds). *Dictionnaire des mouvements sociaux* (pp. 554-564). Paris: Presses de Science Po.

Siméant, J. (2010).La transnationalisation de l'action collective. In E. Agrikoliansky, O. Fillieule, and I. Sommier (eds). *Penser les mouvements sociaux. Conflits sociaux et contestations dans les sociétés contemporaines* (pp. 121-144). Paris: La Découverte.

Smith, J. (2008). *Social Movements for Global Democracy. Baltimore*, MD: The Johns Hopkins University Press.

Smith, M. P. and Guarnizo, L. (eds) (1998). *Transnationalism from Below*. London: Transaction Publishers.

Staggenborg, S. (2010). The Consequences of Professionalization and Formalization in the Pro-Choice Movement. In D. McAdam and D. A. Snow (eds). *Readings on Social Movements. Origins, Dynamics, and Outcomes* (pp. 599-622). Oxford: Oxford University Press.

Svampa, M. (2015). Commodities Consensus: Neoextractivism and Enclosure of the Commons in Latin America. *South Atlantic Quarterly*, 114(1), 65-82.

Swyngedouw, E. (1997). Neither Global Nor Local: 'Glocalization' and the Politics of Scale. In K. Cox (ed.). *Spaces of Globalization: Reasserting the Power of the Local* (pp. 137-166). New York/London: Guilford/Longman.

Tarrow, S and McAdam, D. (2004). 'Scale Shift in Transnational Contention'. In D. Della Porta and S. Tarrow (eds). *Transnational Protest and Global Activism* (pp. 121-147). Lanham, MD: Rowman & Littlefield Publishers.

Vielajus, M. (2009). *La société civile mondiale à l'épreuve du reel*. Paris: Editions Charles Leopold Mayer.

Visseren-Hamakers, I. J., McDermott, C. Vijge, M. J. and Cashore, B. (2012). Trade-offs, Co-benefits and Safeguards: Current Debates on the Breadth of REDD+. *Current Opinion in Environmental Sustainability*, 4(6), 646-653.

Wallbott, L. (2014). Indigenous Peoples in UN REDD+ Negotiations: 'Importing Power' and Lobbying for Rights through Discursive Interplay Management. *Ecology and Society*, 19(1), art. 21. Retrieved from http://www. ecologyandsociety.org/vol19/iss1/art21/

Wilson, J. and Swyngedouw, E. (2014). *The Post-Political and its Discontents: Spaces of Depoliticization*, Specters of Radical Politics. Edinburgh: Edinburgh University Press.

Young, O. R., Berkhout, F., Gallopin, G. C., Janssen, M. A., Ostrom, E. and Leeuw, S. V. D. (2006). The Globalization of Socio-Ecological Systems: An Agenda for Scientific Research. *Global Environmental Change*, 16(3), 304-316.

9

Global Climate Change Finance

SIMONE LUCATELLO
INSTITUTO MORA, MEXICO

Introduction: Environmental aid effectiveness and climate change

Climate change finance and its implementation instruments belong to the broader field of international environmental aid (IEA), which is defined by the OECD as the sum of bilateral and multilateral economic support to developing countries for environmental purposes (OECD, 2012). Over the past 40 years a wide range of responses to environmental problems has been implemented through a set of interacting systems with multiple actors at different scales. Conventional responses at national and global levels include the creation of rules, laws and institutions, with international organisations established to serve as conveners at the global scale. An important pillar of this global strategy for the environment are the economic and financial initiatives. The environment has been high on the agenda ever since the Rio Earth Summit in 1992 and recent focus on the economics of climate change is shifting attention to the costs of climate change mitigation and adaptation for developing countries.

Dominant international literature on global aid effectiveness stresses the point that multilateral aid is preferable to bilateral aid. In order to prove it, most of the studies use empirical methods to draw inferences from highly aggregated cross-national time series data (Abbott and Gartner, 2011). General findings from this literature state that multilateral channels such as the World Bank, the Global Environment Facility (GEF), those controlled by the United Nations (UN) and/or the various regional development banks generally provide greater control to recipient countries. Findings point to the fact that multilateral agencies fund different countries and projects compared to bilateral donors, and multilateral assistance tends to target poorer countries with greater needs (Isenman, 2011). Multilateral aid also tends to be less political, is associated with better outcomes, and appears better able to impose more effective delivery (Martens et al., 2001).

However, there are great challenges in engaging with the multilateral system, which has become increasingly complex. It comprises well in excess of 200 agencies, adding to fragmentation and duplication. While some agreements appear to be high performers (like the Montreal Protocol and to some extent the Kyoto Protocol), the effectiveness of others is seen as limited. Concerns about agreements performance range from perceived institutional complexity, lack of transparency, higher absolute costs and insufficient evidence of effectiveness, among others. With increasing pressure on domestic budgets, donor governments have been placing much greater emphasis on assessing the effectiveness and relevance of different environmental agreements as a guide to how best to distribute both their resources and their staff time between them (Dinham, 2011).

When we narrow down to environmental aid effectiveness (EAE), there is an open debate about how green aid is better delivered multilaterally than to via bilateral channels. The analysis of different databases (like Aid-Data) tells us that environmental aid is increasingly being allocated bilaterally, through national aid agencies, rather than multilaterally, through the international organisations and channels created for this purpose. If we look at historical trends, between 1990 and 2008 the amount of environmental aid channelled through multilateral institutions increased by roughly 16 per cent. In contrast, bilateral environmental aid levels more than doubled over the same period, going from US$3.6 billion to US$6.5 billion. In relative terms, 58 per cent of environmental aid was allocated through multilateral agencies in 1990–94. By 2005–08, this figure had dropped to 42 per cent (RECOM, 2013). Therefore, in contrast with global trends, evidence shows that bilateral green aid is 'preferred' by donors as an effective way to deliver aid, even though further research on specific case studies are required in order to strengthen our understanding of this.

If our understanding of global climate finance is fragmented, the puzzle can be even more complicated when we transfer analysis to regional mechanisms. For example, resources within Latin America and the Caribbean (LAC) region are mostly concentrated in the biggest economies of the region, such as Mexico and Brazil. By contrast, the rest of the countries and in particular those countries which are highly exposed to climate change risk (such as those conforming Central America) have received limited environmental aid so far.

Climate change finance: mechanisms and implementation

International negotiations under the United Nations Framework Convention on Climate Change (UNFCCC) are at a crossroads. At the end of 2015, governments gathered in Paris for the climate summit (COP-21), which successfully framed the new international and universally binding climate agreement. However, to reach this overarching goal requires not only a high political level of ambition but also practical commitments by the international community, particularly in relation to the topic of climate finance. According to recent estimates from COP-21, the public finance offered by developing countries, will result in at least US$18.8 billion per year by 2020 and all multilateral development banks have pledges to scale up climate finance in developing countries substantially by 2020 to more than US$30 billion per year (Nakhooda, 2015).

Many developing countries have stressed the importance of industrialised nations bearing greater responsibility for climate change, given both their historic and current greenhouse gases (GHG) emissions and their superior capacity to respond to climate change. On the contrary, developed nations are most in need of mitigation (especially middle-income countries) and adaptation actions (Brooks et al., 2011). In line with this argument, with the Bali Action Plan in 2007, the UNFCC called for developed nations to provide finance for adaptation and mitigation actions to developing countries.

New and additional resources

Since the Conference of the Parties (COP)-15 in Copenhagen in 2009, and at subsequent COPs (16, 17, 18, 19, 20); developed countries have agreed to provide 'new and additional resources' for adaptation and mitigation. For the long term they have committed to 'a goal of mobilising jointly US$100 billion dollars a year by 2020 to address the needs of developing countries' through a 'wide variety of sources, public and private, bilateral and multilateral, including alternative sources of finance' (OECD, 2011). Efforts to both mitigate and adapt to climate change imply the use of a huge amount of resources, representing a great challenge to the international community and its commitment to support vulnerable countries. However, crucial questions about where this money should come from, who should pay and how and where the money should be delivered are still open for debate. As mitigation and adaptation measures involve a huge financial challenge, it is worth asking who pays for the costs of climate change. How many funds are available to the international community to address the phenomenon? How are these funds distributed and through which mechanisms?

Current principles governing climate change and the use of international funds are based on the provision of international financial aid to climate change mitigation and/or adaptation, which should be seen as additional to development assistance. This has been a central element of the international climate change agreements from the outset (Falkner, 2013). In fact, the UNFCCC – agreed in 1992 – stated that developed countries shall provide new and additional financial resources to developing countries. As Ivanova points out (2013), expanding the donor base, increasing funds availability and ensuring predictable and stable financial flows are currently the main priorities of international environmental governance.

The current international financial architecture for climate change is made up of three main sources. One is bilateral, which comes from direct cooperation between governments and executed through direct transfers from developed to developing countries. The second type of source is multilateral, focusing on climate investment funds and multilateral organisations like the World Bank and regional multilateral banks. A third source is the set of mechanisms established by the UNFCCC, where governance processes of the funds and their implications have greater legitimacy under the regime of the Convention. These mechanisms include the Global Environment Facility (GEF), the Adaptation Fund (AF), the Climate Investment Fund (CIF) and, most recently, the Green Climate Fund (GCF), as well as new financial mechanisms such as results-based payments for reducing emissions from deforestation, degradation, forest conservation (REDD+) and clean energies (CEMDA, 2013). In addition, developing countries have increased their own spending, through their own national budgets, on activities related to climate change. However, the multidimensional and cross-cutting nature of climate change suggests that amounts of international public finance are still meagre given the magnitude of the phenomenon.

Currently, there are more than 50 international public funds, 60 carbon markets (formal and voluntary) and 6,000 private investment funds, which support so-called 'green' funding. In addition, multiple types of financing (such as Carbon 1, financing for REDD+, etc.) and a variety of tools for delivery and financing packaging (such as results-based sectoral approaches, payments, etc.) are rapidly emerging and evolving while posing additional new challenges. In the case of Mexico in 2012, the country – along with ten others – absorbed almost 45 per cent of the international resources for climate change (CEMDA, 2013). In addition, a growing number of recipient countries have set up national climate change funds that receive funding from multiple developed countries in an effort to coordinate and align donor interests with national priorities. Global climate finance architecture is therefore a complex matter.

As for the release of funds, the means currently available are mechanisms that are private and public in nature. These include grants, concessional loans, and equity and delivery mechanisms for resource-based projects under the framework of clean development mechanisms (CDM).

Challenges

In critical terms, the current proliferation of climate finance mechanisms increases the challenges of coordinating and accessing finance (Fankhauser and Burton, 2011). As mentioned above, climate finance involves flows of funds from developed to developing nations to help poorer countries to cut their emissions and adapt to the impacts of climate change. US$356 million has been pledged and US$749 million deposited to these funds since last year. The largest contributors to these funds were the United Kingdom (UK), the United States of America (USA), Germany and Japan. Between October 2012 and September 2013, US$431 million were approved for new projects and US$429 million disbursed to support 157 projects, a 23 per cent increase from the number of projects approved the previous year (Climate Funds Update, 2015).

In 2013, annual global climate finance flows totalled approximately US$331 billion, falling US$28 billion below 2012 levels. Public actors and intermediaries contributed US$137 billion, which was largely unchanged from the previous year. Private investment totalled US$193 billion, falling by US$31 billion or 14 per cent from 2012. Climate finance flows were split almost equally between developed (OECD) and developing (non-OECD) countries, US$164 billion and US$165 billion respectively. The amount tracked flowing from developed to developing countries fell by US$8 billion from 2012, to US$34 billion, with multilateral contributions from development finance institutions (DFI) falling by US$5 billion, and private investment contracting by US$2 billion. Almost three-quarters of total flows were invested in countries of origin. Private actors had an especially strong domestic investment focus with US$174 billion or 90 per cent of their investments remaining in the country of origin. This demonstrates that investment environments that are more familiar and perceived to be less risky are essential to influence investment decisions, highlighting the importance of domestic policy frameworks in unlocking scaled up climate finance flows.

Performance at regional level presents different features. Latin American countries, for example, have performed differently. According to the Economic Commission for Latin America and the Caribbean (ECLAC); since 2004 through 2012, US$2.035 billion have been approved for 220 projects in the region. This amount increased 118 per cent over 2011. Of this total,

US$1.143 billion took the form of grants supporting the majority of approved projects. US$892 million are provided in the form of concessional loans for 10 projects financed by the Clean Technology Fund (CTF) and one Forest Investment Program (FIP) supported project under the World Bank's Climate Investment Funds (CIFs), which are implemented in the region by the Inter-American Development Bank (IADB). As of October 2012, the total amount disbursed was US$397.15 million for 110 projects (Climate Funds Update, 2015).

> Most of this finance is provided as concessional loans. The Global Environment Facility has disbursed the largest volume of finance to the region to date: approximately US$169 million in grants for 44 mitigation projects. The United Kingdom, Norway, Japan and Germany, are also investing in LAC. Japan and Norway are the largest bilateral contributors, with Japan providing US$347 million mainly for mitigation in the private sector, while Norway has provided more than US$337 million for two programs that support REDD+ (Cabral y Bowling, 2014).

However, part of this economic puzzle, is to understand if and how these funds are attending the most urgent climate problems in the region, such as the retreat of glaciers, which could lead to water stress for around 77 million people by 2020, and continued deforestation of tropical forests. Latin America and the Caribbean's vulnerability to the likely impacts of climate change, exacerbated by persistent income inequality and poverty, means that adaptation needs in the region must become more central to national sustainable development strategies.

Green Climate Fund

An important attempt to put order to this complex puzzle has been the creation of a new instrument, the Green Climate Fund (GCF), which entered into force in December 2013. Through this mechanism, developing countries are keen to get financing without going through international institutions, like the World Bank, and being subjected to their rules and conditions. The GCF plans to provide US$100 billion annually from 2020 to support mitigation and adaptation to address the problem. The GCF was proposed in 2009 during the Conference of the Parties of UNFCCC in Copenhagen (Denmark), and approved the following year at meetings of the COP in Cancun (Mexico, 2010). Mexico was one of the Fund's principal global promoters. Potential beneficiaries of this fund are mainly countries running transportation projects aimed to reduce carbon dioxide emissions but also include relocation efforts of communities affected by rising sea levels, drought and crop damage, and a long list of other projects. The Fund is governed by a board of 24 members with equal representation of developed and developing countries. Among the members are representatives of Chile, Peru, Colombia, Mexico, Cuba and Belize, who share the chair. The World Bank is the administrator of the interim fund during the first three years.

In August 2012, the Fund's board met for the first time in Geneva, where they began to lay the foundations for operation. In this first meeting, a list of issues and pending tasks were drafted but no real progress was made in terms of effectiveness to implement operating mechanisms of the GCF. A second meeting of the board was held in Songdo, South Korea, in October of the same year and discussions began on the rules relating to the participation of observer countries. However, as on the previous occasion, the board did not fundamentally advance in the planning of activities (Lattanzio, 2014).

In 2013, the GCF board of directors had their third meeting in Berlin, at which various decisions were made. Notable among them were the 'additional rules of procedure', a set of actions for the operation of the GCF. It also details the participation and the role of civil society observers in the Fund, as well as the process for accreditation. The board met again in June 2013 in South Korea and in October in Paris. At these meetings progress was made in strengthening the operational structure of the Fund and demarcating the necessary tasks for prompt operation. Despite its slow development, in December 2013, the GCF was inaugurated at its headquarters in Seoul, South Korea. The launch was symbolic, since the Fund would not be fully operational until the second half of the following year.

At this stage, the Fund's finances are rather fragile. Several donor countries that had pledged to provide funds have not contributed as planned. In 2010, rich nations pledged to provide US$10 billion a year between 2011 and 2013, and raise funds to US$100 billion annually by 2020. Yet the influx of money has been much less – even falling by two thirds in 2013 compared to 2012. The Fund currently has only US$40 million available; a promised sum by South Korea must also cover the administrative costs of the new headquarters. An important issue in creating the Fund is that resources will come from not only public transfers but also private investment. The idea is that innovative sources of financing supplement classical budgetary resources to support the Fund.

Nevertheless, the formal launch of the GCF involves huge challenges and the possible outcomes and impacts for recipients still await a verdict. First, the implementation of the Fund is still in the hands of the board and the

UNFCCC. The GCF is designed as an executive instrument of the UNFCCC with an 'independent' board, a general secretariat, etc. Negotiations in Durban concluded that the Fund would operate under the guidance of the COP of member countries and, unlike other funds (such as that of Adaptation), provide work under the guidance and the 'authority' of the COP. Such a subtle distinction signified profound differences for negotiators from China and the G77; they see in the second option (based on the model adaptation fund) a proposal which may leave decisions on the composition of the board and, especially, the use of funds in the hands of the countries of the global North (Lucatello, 2014).

Conclusion

Within the broader context of international environmental aid, climate change finance plays an increasing role both in terms of gathering new financial support from donor countries directed to developing ones and in the number of actors and financial schemes available. There have been various initiatives, starting with international mechanisms such as the UNFCCC, the Kyoto Protocol (which expired in 2012) and the Montreal Protocol. Alongside regulatory efforts, the international community has driven transformation processes to provide countries with policies and technologies that can catalyse new investments, inserting climate change into existing national systems. Additionally, these efforts should provide significant support to build resilient systems, particularly for the poorest and most vulnerable developing countries, which are those that have contributed least to the accumulation of GHG in the atmosphere.

The available global funding and capacity to absorb these resources vary according to donor agencies, countries and private flows available. While developed countries have internal capabilities to generate and utilise climate funding, many developing countries lack the resources, skills or institutional systems and policies to use climate funding effectively. Such barriers are accentuated in countries with large vulnerable groups, such as the poor and women, thus threatening the attainment of the goals of poverty reduction and the future Sustainable Development Goals (SDGs). Moreover, major financial investments – from both public and private sources – are also required to transition national economies to a low-carbon path, reduce greenhouse gas concentrations to safe levels, and build the resilience of vulnerable countries to climate change.

However, as seen before, challenges to creating robust climate change finance are substantial: in developing countries, direct government funding is scarce and international environmental aid is becoming less concentrated.

Pledges by international donors remain inadequate for the magnitude of the challenge of stabilising a steep trajectory of greenhouse gases. Additional financial investments must accompany national efforts to mitigate climate change effects, although rules, regulations and fiscal incentives ought to be strongly promoted by those who are in greater need for climate financial support. In this complex panorama, the global financial architecture for climate change is an evolving issue, where actors and rules are constantly engaging in transformation.

References

Abbott, K. W. and Gartner, D. (2011). *The Green Climate Fund and the Future of Environmental Governance*. Earth System Governance Working Paper No. 16. Lund and Amsterdam: Earth System Governance Project. Retrieved from http://www.ieg.earthsystemgovernance.org/sites/default/files/files/publications/ESG-WorkingPaper-16_Abbott%20and%20Gartner.pdf

CEMDA (2013). *La arquitectura financiera para el cambio climático en México. Retos y propuestas para una política financiera transparente y eficiente para la mitigación y adaptación al cambio climático en México.* [Finance Architecture for Climate Change in Mexico. Challenges and Proposals for a Transparent and Efficient Finance Policy for Climate Change Mitigation and Adaptation in Mexico]. Retrieved from http://financiamientoclimatico.mx/recursos/FCM.pdf

Falkner, R. (ed.) (2013). *The Handbook of Global Climate and Environment Policy*. Chichester: Wiley Blackwell.

Brooks, N., Anderson, S., Ayers, J., Burton, I. and Tellam, I. (2011). *Tracking Adaptation and Measuring Development*. IIED Climate Change Working Paper No. 1. London: International Institute for Environment and Development. Retrieved from http://pubs.iied.org/10031IIED.html

Cabral y Bowling, R. B. (2014). Fuentes de financiamiento para el cambio climático. [Sources to Climate Change Finance]. No. 254. CEPAL. Retrieved from http://repositorio.cepal.org/bitstream/handle/11362/37217/S1420542_es.pdf?sequence=1

Climate Funds Update (2015). *Latest Climate Fund Data*. Retrieved from http://www.climatefundsupdate.org/

Dinham, M. (2011). *Study of AusAID's Approach to Assessing Multilateral Effectiveness. A Study Commissioned by the Independent Review of Aid Effectiveness to Assist in Their Overall Analysis of the Effectiveness and Efficiency of the Australian Aid Program.* Sydney. Retrieved from www.alnap. org/pool/files/study-multilateral.pdf

Fankhauser, S. and Burton, I. (2011). *Spending Adaptation Money Wisely.* Centre for Climate Change Economics and Policy Working Paper No. 47. London: The Grantham Foundation/ESRC/University of Leeds/LSE.

Ivanova, M. (2013). Reforming the Institutional Framework for Environment and Sustainable Development: Rio+20's Subtle but Significant Impact. *International Journal of Technology Management and Sustainable Development,* 12(3), 211-231.

Isenman, P. (2011). *Architecture, Allocations, Effectiveness and Governance: Lessons from Global Funds.* ODI Meeting on Climate Change. London: Overseas Development Institute.

Lattanzio, R. (2014). *International Climate Change Financing: The Green Climate Fund (GCF).* Washington, DC: Congressional Research Service. Retrieved from https://fas.org/sgp/crs/misc/R41889.pdf

Lucatello, S. (2014). Global Financial Architecture for Climate Change and the Green Fund. In: L. Lázaro Rüther, C. Ayala Martínez, and U. Müller (Eds). Global Funds and Networks. *Narrowing the Gap between Global Policies and National Implementation* (pp. 190-197). Eschborn, Germany/Mexico City, Mexico: GIZ-Nomos-Instituto Mora.

Martens, B., Mummert, U., Murrell, P. and Seabright, P. (2001). *The Institutional Economics of Foreign Aid.* Cambridge: Cambridge University Press. Retrieved from http://ostromworkshop.indiana.edu/papers/ martens040901.pdf

Nakhooda, S. (2015, December 12). *Climate Finance: What Was Actually Agreed in Paris?* Overseas Development Institute. Retrieved from http://www. odi.org/comment/10201-climate-finance-agreed-paris-cop21

OECD (2011). *Busan Partnership for Effective Development Cooperation: Statement of the Fourth High Level Forum on Aid Effectiveness.* Busan, South Korea: OECD. Retrieved from http://www.oecd.org/dac/ effectiveness/49650173.pdf

OECD (2012). *Development Perspectives for a Post-2012 Climate Financing Architecture*. Paris: OECD. Retrieved from http://www.oecd.org/greengrowth/green-development/47115936.pdf

RECOM (2013). *Trends in Environmental Aid: Global Issues, Bilateral Delivery*. Research and Communication on Foreign Aid. Retrieved from http://recom.wider.unu.edu/article/trends-environmental-aid-global-issues-bilateral-delivery

SECTION III:

TWO STEPS FORWARD, ONE STEP BACK: PERSPECTIVES AS WE CONTINUE WITH OUR LIVES

10

New Practices and Narratives of Environmental Diplomacy

LAU ØFJORD BLAXEKJÆR
UNIVERSITY OF THE FAROE ISLANDS

Introduction: governance, diplomacy and IR

Why should we pay attention to environmental diplomacy in International Relations? And how should we seek to understand and explain environmental diplomacy? In the following chapter, I begin by outlining some of the challenges facing diplomacy and diplomatic studies together with some recent theoretical responses to these challenges. Focusing on two examples of important changes in global environmental governance, this chapter argues that environmental diplomacy plays a significant role and that practice theory and narrative theory offer better models of analysis than more mainstream institutionalist, regime theory, multilevel governance or discourse theory approaches.

Narratives are seen as discursive constructs but also as constructed by and constructive of the practices through which environmental governance is pursued. The two chosen phenomena of the United Nations Framework Convention on Climate Change (UNFCCC) negotiations and green growth governance are central today and will likely continue to be so as the world negotiates the transition to an environment- and climate-friendly economy. Some are optimistic in relation to the impact of renewable energy, finding that 'the question is no longer if the world will transition to cleaner energy, but how long it will take' (Randall, 2015). Others might be more realistic when describing the political challenges, stating that 'it is clear that if the world is to move towards a significantly more carbon-efficient and climate-resilient pathway of economic growth, a much more compelling economic case for action has to be made' (The New Climate Economy, 2015). Whichever of the above routes is subscribed to, the world clearly appears to be in a transitional phase. In practice, environmental diplomacy offers considerable insight for International Relations to understand and explain this transformation.

Diplomacy: challenges and new approaches

Diplomacy is hardly a new topic in IR, and several works have been written in relation to different aspects of diplomacy, such as the dialogue between states and how Western-style diplomacy has come to dominate world politics (Watson, 1982; Bull and Watson, 1984), the United Nations (UN) conference diplomacy (Kaufmann, 1988) or environmental diplomacy (Benedick, 1998; Chasek, 2001; Susskind et al., 2014). Nor is it new that diplomacy and diplomatic studies are seen as irrelevant, criticised as dangerous (Wiseman, 2011: 710) or characterised as the poor child of IR (Pouliot and Cornut, 2015: 297). While they may have been written off in the face of globalisation and increased nationalism, diplomacy and diplomatic studies are alive and kicking, as exemplified by the recent *Oxford Handbook of Modern Diplomacy* (Cooper et al., 2013), the four-volume set *International Diplomacy* (Neumann and Leira, 2013) and *Diplomacy and the Making of World Politics* (Sending et al., 2015). The classic understanding of diplomacy is that of negotiation, persuasion and dialogue between equal and sovereign states performed by a highly educated corps of Ministry of Foreign Affairs (MFA) civil servants/ diplomats (see Kaufmann, 1988). This image is being challenged in many ways. Here, I highlight just three: 1) diplomacy is also practiced by non-MFA civil servants and non-state actors (e.g. Neumann, 2002; Cooper et al., 2013); 2) diplomacy as practice takes place according to an international pecking order or social field, in which some states and diplomats are higher up the hierarchy than others – power in practice is relational (Adler-Nissen and Pouliot, 2014); and 3) there is an ontological discrepancy between the world of diplomats and mainstream IR scholarship (Adler-Nissen, 2015) to the effect that, for example, 'American IR *theory* now lags behind American diplomatic *practice*' (Wiseman, 2011: 711). The classic image of diplomacy has been replaced by an image of hybrid diplomacy with multiple actors, multiple issues and multiple practices.

Before pointing out the recent theoretical responses, this chapter will outline two specific examples of how global environmental governance and diplomacy have been challenged. These two examples spring from the general understanding that the United Nations (UN) system seems incapable of reaching effective solutions to environmental and climate change issues, especially following the breakdown of climate negotiations at the Conference of the Parties (COP)-15 in Copenhagen, Denmark, 2009. 'The general view [...] was that there was something inherently dysfunctional about climate diplomacy under the UNFCCC in particular and under the UN in general' (Elliot, 2013: 848). A diplomat and scholar further noticed that:

> The current situation in the UN multilateral process is worse than before Copenhagen. Failure to adopt the lightest possible nonbinding declaration underscores the bleak prospects of the consensus-based UN process for responding to climate change. [...] Whereas previously we held out hope that omnipotent heads of state could resolve outstanding political differences in one fell swoop, we have now lost even that hope. If the highest-level leaders cannot settle differences, who can? (Dimitrov, 2010: 22)

Leading up to COP-16, 'key players [said] a further breakdown in fresh discussions [...] could spell the end of the UN multilateral negotiating process' (Willis, 2010), leading to climate minilateralism in fora such as G8, G20 or maybe only the G2 (the United States and China) (Casey-Lefkoeitz, 2010).

The COP-15 setback should be understood in relation to concurrent trends relating to the increasing institutional fragmentation of global environmental governance (Zelli and Van Asselt, 2013), climate experimentation (Hoffmann, 2011) and transnational climate change governance arrangements (Bulkeley et al., 2014). Despite these tendencies and new spaces of climate governance, UNFCCC negotiations got back on track. Both within and outside the UNFCCC, we see new diplomatic communities and global partnerships being formed to address the challenges of a needed transition to a new economic system. These two general fields of global environmental governance also overlap.

The UNFCCC response

COP-16 reached a consensus understanding, adopted the Cancun Agreements and re-established some trust and legitimacy in the UNFCCC. COP-17 in Durban, South Africa, delivered a text known as the Durban Platform. COP-21 in Paris, France, has delivered a new global agreement in 2015. COP-18 in Doha, Qatar, managed to finish the old negotiations from COP-13 (the Bali Action Plan) and set a more detailed timetable for the Durban Platform (Christoff and Eckersley, 2013: 118; UNFCCC, 2014a; 2014b). COP-18 also formally adopted a second commitment period under the Kyoto Protocol. In a recent analysis based on a large number of interviews with negotiators, it is argued that the diplomatic efforts of the Mexican COP presidency, which was anchored in the MFA, and informal dialogues such as the Cartagena Dialogue for Progressive Action made the difference in getting negotiations back on track (Monheim, 2015: 29–62). Blaxekjær and Nielsen (2014) have further demonstrated that new political groups such as the Cartagena Dialogue have changed the narrative landscape of negotiations, thus creating a new possible space of constructive negotiations.

Green growth networks

Korea and Denmark, together with COP-16 host Mexico, began promoting green growth in connection with UNFCCC events as a supplementary process to the UNFCCC negotiations, and they established the Green Growth Alliance in 2010. This alliance was joined by China, Kenya and Qatar in 2012, Ethiopia in 2014 and Vietnam in 2015. Korea and Denmark have been shaping green growth to encompass public–private partnerships through investments in the green sector and market-driven principles but have still actively connected green growth in order to contribute to global sustainable development and poverty eradication. One purpose besides fighting climate change seems to be to position Korea, Denmark and Mexico as responsible members of the global community – with the bridging of the North–South divide as a central goal of the Green Growth Alliance.

In 2010, with Danish financial support, Korea established the Global Green Growth Institute (GGGI), and Denmark initiated the Global Green Growth Forum (3GF). Since 2012, GGGI has been a full-fledged international organisation, operating as a forum for partners and as the secretariat of the Green Growth Alliance placed in Denmark's Ministry of Foreign Affairs. Both 3GF and GGGI have actively promoted green growth at their own summits and to other platforms, such as the UNFCCC, G20 and Rio+20 summits.

The IR responses to changes in global environmental governance

The Cartagena Dialogue and 3GF exemplify important changes in global environmental governance in which diplomacy plays a critical role. These examples could be analysed with the dominant global environmental governance approaches, regime theory or multi-level governance theory (Okereke and Bulkeley, 2007). The choice of one over the other will skew the analysis, however, as some theoretical/empirical problems are yet to be resolved. Regime theory is skewed towards states and material interests at the international level, missing the importance and independent role of non-state actors, ideas and governance on multiple levels. Multi-level governance theory takes the opposite stance. Both seem to miss the critical role played by state and non-state diplomats. The climate negotiations literature generally falls within the rationalist, constructivist and descriptive approaches (Dimitrov, 2013: 340). The first two are often theoretical, whereas the latter is often policy-oriented. The rationalist or interest-based approaches often take the shape of hypotheses, which shape our understanding of negotiations as dichotomous: 'agreement/no agreement, action/no action, or cooperation/ conflict' as the only options (Bernstein, 2001: 10).

> Much published work offers recycled information that can be derived without negotiations actually having been observed. [...] The dynamics around the negotiation table often remain hidden. What is the verbal exchange? What are the offers and responses made during informal consultations? Relevant literature tends to avoid these questions and gravitate toward related topics such as theorizing about the creation of institutions and their impact on state behavior.(Dimitrov, 2013: 346–347)

IR has recognised the gap between meso-level institutions and micro-level practices to some extent, especially in relation to issues of trust and cooperation (Farrell, 2009; Walker and Ostrom, 2009). Further, Holmes (2013: 829) argues that personal relationships in negotiations can promote cooperation, because '[f]ace-to-face meetings allow individuals to transmit information and empathize with each other, thereby reducing uncertainty, even when they have strong incentives to distrust the other'. The IR literature is on its way to be both theoretically informed and empirically strong and thus to pay closer attention to the diplomats and governance practices (Eckersley, 2012; Audet, 2013; Bauer, 2013: 332; Dimitrov, 2013: 346).

The international practices approach

A new approach receiving much attention within IR is that of international practices. Wiseman (2011; 2015), Brown (2012) and Adler-Nissen (2013) point to Adler and Pouliot (2011) as the central work of influence, although the roots of the practice can be traced further back. The broader (Bourdieusian) research agenda focusing on international practices such as diplomacy has proved very useful and innovative in bridging dichotomies and explaining social change (Neumann, 2002; Bigo, 2005; Adler, 2008; Pouliot, 2008; Adler and Pouliot, 2011; Bigo and Madsen, 2011; Adler-Nissen, 2013; Adler-Nissen and Pouliot, 2014). According to Adler and Pouliot:

> [W]hile we agree that practices have an epistemic or discursive dimension, we broaden practices' ontology and thus do not limit the scope of our study to text and meaning.

> Rather, practice forces us to engage with the relationship between agency and the social and natural environments, with both material and discursive factors, and with the simultaneous processes of stability and change. In fact, the concept of practice is valuable precisely because it also takes us outside of the text. (Adler and Pouliot, 2011: 2–3)

> Practices are socially meaningful actions that reify background knowledge and discourse in and on the material world. Practices are the dynamic material and ideational processes that enable structures to be stable or to evolve, and agents to reproduce or transform structures. (Adler and Pouliot, 2011: 6)

Research and researchers must engage in and interpret the field. It builds on what Wagenaar calls dialogic meaning in action, where '[i]t doesn't make sense to try to locate meaning ontologically in the mind or in some reified cultural or institutional pattern' (Wagenaar, 2011: 21). Researchers should approach meaning through the study of social action as something that is both particular to the specific actor and moment and also generally meaningful, because it signifies something larger. As such, this understanding of international practices fits well with the new hybrid image of diplomacy. It is also argued that diplomatic studies and practice theory can learn a lot from each other (Pouliot and Cornut, 2015). Furthermore, this approach can benefit from insights from a narrative approach as a way of including and analysing strategic rhetoric, language and discourse to take us outside the text but still keep texts as important sources and as examples of practices, especially in negotiations and diplomacy (Blaxekjær and Nielsen, 2014).

> [Meaning] is more a shared set of understandings that are linguistically and actionably inscribed in the world, and that are invoked, and, in an ongoing dialectical movement, sustained, whenever actors engage in a particular behavior, and whenever we 'read' the symbolic meaning of that particular behavior. (Wagenaar, 2011: 21).

Gaining access to practices is the first task to be overcome by the researcher. This is not always possible due to practical barriers and the secret nature of diplomacy (although the hybrid version is less secret). However, 'even when practices cannot be "seen", they may be "talked about" through interviews or "read" thanks to textual analysis' (Pouliot, 2013: 49). In the following, I turn to a brief analysis of the Cartagena Dialogue and the 3GF as sites of environmental diplomacy.

The Cartagena Dialogue for Progressive Action as a diplomatic community of practice

Diplomacy in the form of dialogue is hardly a new phenomenon in UNFCCC negotiations. Examples include the Greenland Dialogue on Climate Change, which existed from 2005 to 2009 and was launched by the Danish COP-15 presidency (Meilstrup, 2010: 120); the Petersberg Dialogue, co-hosted by Germany and Mexico in 2009; the Geneva Dialogue on Climate Finance, co-hosted by Mexico and Switzerland in 2009 (Monheim, 2015: 50–51); and the more recent Toward 2015 Dialogue, convened by the Centre for Climate and Energy Solutions (C2ES, 2015). These discussions can best be described as high-level meetings preparing for a specific COP. Yet, the Cartagena Dialogue is different and better described as a political group or community that works in and between COPs (Blaxekjær and Nielsen, 2014).

The Cartagena Dialogue was officially formed in March 2010 in Cartagena, Colombia, by UNFCCC negotiators and experts from around 30 parties representing all regions. It did not suddenly emerge; many experienced negotiators from the European Union (EU), the Alliance of Small Island States (AOSIS), the Least Developed Countries (LDCs) and Latin America and the Caribbean (LAC) had cultivated an informal community for many years, but it was the common experience of failure at COP-15 and the feeling of being excluded from influence when USA and BASIC (i.e. Brazil, South Africa, India and China) negotiated the Copenhagen Accord that brought negotiators together more formally. The founding meeting resulted in an understanding of the community as 'an informal space, open to countries working towards an ambitious, comprehensive and legally binding regime in the UNFCCC, and committed domestically to becoming or remaining low-carbon economies' (as reported by Lynas, 2011, who worked for the Maldives at the time; corresponding to UNFCCC webpage; confirmed by own interviews). Participants agreed to be explicit about the Cartagena Dialogue not being described as a political group but rather as a dialogue with participants instead of members (as pointed out by all of the interviewees).

Following the international practices approach and paying attention to new narratives, I have previously analysed the Cartagena Dialogue as a community of practice or diplomatic community under the UNFCCC (Blaxekjær, 2015; see also Blaxekjær and Nielsen, 2014). A community of (diplomatic) practice is broadly defined as 'groups of people who share a concern, a set of problems, or a passion about a topic, and who deepen their knowledge and expertise in this area by interacting on an ongoing basis' (Wenger et al., 2002: 4). The community of practice concept is further developed in IR:

> The community of practice concept encompasses not only the conscious and discursive dimensions, and the actual doing of social change, but also the social space where structure and agency overlap and where knowledge, power, and community intersect. Communities of practice are intersubjective social structures that constitute the normative and epistemic ground for action, but they also are agents, made up of real people, who – working via network channels, across national borders, across organizational divides, and in the halls of government – affect political, economic, and social events. (Adler and Pouliot, 2011: 17–18)

Common to communities of practice is that they consist of a basic structure defined as 1) a community of people who care about an issue, 2) a domain of knowledge, which defines a set of issues, and 3) the shared practices that they are developing to be effective in their domain. To identify a community of practice, one must go beyond the abstract and desk-based studies: 'You have to look at how the group functions and how it combines all three elements of domain, community, and practice' (Wenger et al., 2002: 44).

The Cartagena Dialogue illustrates a major change in global environmental governance, because it is a new kind of diplomatic community bridging North and South positions by redefining the longstanding, dividing principle of Common But Differentiated Responsibility (CBDR). Through collaborative and dialogic work beginning with the common interest in understanding the views of other parties, building further trust at the personal level and then working together to explore possible common ground on specific negotiation issues, the Cartagena Dialogue played an important role in getting climate negotiations back on track after the breakdown at COP-15 (Blaxekjær, 2015). The Cartagena Dialogue is a community that thrives on boundary spanning as a defining practice whereby necessary face-to-face dialogue between North and South and between political groups actually takes place; necessary, because it creates the basis for trust, learning and new ideas, which contribute to negotiations moving forward. Participants in the Cartagena Dialogue have changed their normal negotiation practices. They now find it easier and are more comfortable communicating with one another if/when they need something clarified. This increased level of frank communication across the North–South divide is an important change.

Text is central to climate negotiations. And although the Cartagena Dialogue does not produce official statements and submissions, their new diplomatic practices include taking notes of common ground on various issues.

This common language finds its way into the submissions and statements of the respective delegations. Both the Mexican and South African COP presidencies were able to use the Cartagena Dialogue as a sounding board for difficult issues. After the failure of COP-15, where especially the EU, AOSIS and LDC felt they were excluded from influence, the Cartagena Dialogue has become a community of practice, where parties supporting a global, action-oriented approach can engage with one another and develop compromises based on in-depth knowledge of the other parties' positions and reasoning. Such compromises have the potential to be strong and long-lasting and put pressure on laggards. These practices and their organisation in a community are part of a new narrative in UNFCCC negotiations, where CBDR is reinterpreted to also mean that developing and developed countries alike should take as ambitious action as possible, including mitigation (for full analysis, see Blaxekjær, 2015).

Green growth governance and hybrid diplomacy

Previous research by the author (Blaxekjær, 2015) has highlighted how green growth is being carved out as a new global environmental governance subfield, which confirms three trends in global environmental governance: a) the increasing institutional fragmentation of global environmental governance (Zelli and Van Asselt, 2013); b) that activities, initiatives and networks fit under a broad neoliberal discourse (Bernstein, 2001; Bulkeley et al., 2014); and c) that all regions of the world experience the emergence of new partnerships that cross North and South and link public and private actors (Bulkeley et al., 2014; Blaxekjær, 2015). I also observe that orchestration – or governmentality as the conduct of conduct in governance and discursive terms – is an important part of this hybrid diplomacy. It is important to study and understand the orchestrating role played by diplomacy in these new, overlapping environmental governance fields if we want to be able to explain how global politics unfolds. The concept of hybrid diplomacy is receiving increasing attention; however, a Google Scholar search reveals that IR has not developed a theory of hybrid diplomacy as such. As Sending et al. (2015: 1) note, 'as we enter the twenty-first century, everybody seems to agree that diplomacy is changing, yet few people can specify exactly how – and with what effects on world politics'.

From a practice perspective, I understand hybrid diplomacy as a category of practice which, on the one hand, includes and mixes not just different issues (e.g. political, economic, environmental) and different actors (professional diplomats, other civil servants, and non-state actors) but also different practices; that is, expanding our understanding of what diplomacy is in practice. On the other hand, I understand hybrid diplomacy as a category of analysis with three dimensions: 'first, diplomacy is a process (of claiming authority and jurisdiction); second, it is relational (it operates at the interface between one's polity and that of others); and third, it is political (involving both representation and governing)' (Sending et al., 2015: 6). In this sense, it is advisable to analyse political and governing processes and relations as constituent of 'how diplomacy is involved in generating agents (e.g. states), objects (e.g. treaties, embassies), and structures (sovereignty)' (Sending et al., 2015: 7). Thus, the term hybrid refers to both the first category of practice and the category of analysis, the latter being a mix of especially practice theory approaches (Pouliot and Cornut, 2015), and also the fact of being analysed by diplomats, scholars and scholar-diplomats.

Green growth governance as an empirical phenomenon lives up to the above definitions, because when there is contact between polities, there is diplomacy at work. In Table 1 below, I have listed most of these new specific green growth networks established in the period 2005–2013. All networks have international participants. These networks are the new spaces and places where global environmental governance – governing and governmentality – is forged through diplomacy.

Table 1: Examples of international green growth networks

Network Name	Established by (year)
Seoul Initiative Network on Green Growth	Korea (2005)
ASEM Green growth Network	Korea (2008)
Global Green Growth Institute (GGGI)	Korea (2010)
Global Green Growth Forum (3GF)	Denmark, Korea and Mexico (2010)
Nordic Council of Minister's green growth working group	Denmark (2010)
Green Growth Cities	OECD (2010)
Astana Green Bridge Initiative (AGBI)	Kazakhstan and UNESCAP (2010)
Green Growth Initiative	African Development Bank and OECD (2011)
STRING green growth network	STRING members (2011)
Myanmar Green Economy Green Growth	GEGG (private and academic partners) (2011)
Mekong region's green growth network	Vietnam, ADB, UK, UNEP, WWF, Denmark (2011)

Green Growth Knowledge Platform (GGKP)	KOREA, OECD, UNEP and World Bank (2012)
C40 green growth network	New York and Copenhagen (2012)
G2A2 – Green Growth Action Alliance	B20 Task Force on Green Growth (2012)
Green Growth Best Practice (GGBP)	GGGI, Climate Development and Knowledge Network (CDKN), the European Climate Foundation, and the International Climate Initiative (German Government). GGGI serves as the executive agency (2012).
Green Growth Group	United Kingdom (2013)

Source: Adapted from Blaxekjær (2015).

Through GGA and 3GF we can observe how diplomacy has become hybrid in relation to issues, participating actors, and their practices. The GGA is an unusual international alliance between sovereign states; countries that we would expect to cooperate bilaterally but not in a minilateral alliance, because they are very diverse and belong to very different groups in political, economic, cultural and climate terms. These are not countries one would expect to work together in the UNFCCC, but the GGA works for the same kinds of goals as the UNFCCC. The 3GF secretariat is hosted by Denmark under the auspices of the Ministry of Foreign Affairs and run by career diplomats. The GGA seeks to foster and develop global public–private partnerships capable of contributing to the transformation to a global green economy. This work is strategically connected with global and regional processes such as the UNFCCC, SDGs or different water and energy programmes. There is a growing literature on green growth (for a review, see Blaxekjær, 2015), but it has primarily overlooked the new hybrid diplomacy at play and has not asked why, for example, actors who are in different UNFCCC groupings and oftentimes in opposition are suddenly working together in the GGA. The 3GF annual conference was first held in 2011, and the first three conferences were attended by heads of state, UN secretary-general Ban Ki-moon, royalty, CEOs, directors from OECD, the World Bank and UN agencies, as well as financial institutions, and a few researchers and NGOs. Although these first-tier leaders participated with engagement and resolution, the mode has shifted since 2014 towards getting the second-tier leaders together to get closer to the implementation of new partnerships and work plans. 3GF is also being repositioned to support and facilitate the implementation of COP-21 and SDGs.

On a personal note: as a participant observer, I attended the annual meetings in 2011 and 2014 and some preparatory meetings in 2014 and 2015. I have interviewed the secretariat and several participants. My findings in relation to the role of hybrid diplomacy are that, first, there is a division of labour between the GGA's MFA diplomats and other participants. Although there are a multitude of actors participating, the MFA diplomats assume the role as facilitators and orchestrators. As such, they are closer to setting the rules of the game than others – although these rules must be broadly accepted by participants and recognised as competent performances, which they are. Non-MFA diplomats also play important roles by helping to prepare the meetings and develop and utilise personal networks, competencies and experiences from, for example, being part of the top management of a global energy or beverage company, global consultancy firm, international organisation or international non-governmental organisation. This all leads to getting the new partnerships to the operating stage and, thus, assist in the implementation of a new governance structure for an environmental political economy. As stated on the 3GF homepage:

> Collaborative partnerships are the cornerstone of 3GF's work. They are considered a key enabler to accelerate the transition to an inclusive green economy. [Twenty-three] partnerships are currently working fully under the 3GF umbrella, while several more participate in the 3GF process but may not have evolved past an initial conversation. (3GF, 2015)

Issues are also multiple, ranging from mitigation, adaptation, water management, information, urban planning and development, agriculture, energy, transport, and more. The practices are multiple: the traditional practices in international politics, such as conferences, meetings in own and other countries (e.g. 3GF has held other meetings in Chile, Colombia, China and Kenya), MOUs, strategic communication, networking, displaying technological solutions, and cultural packaging with dinners and concerts. The non-traditional practices include a strong focus on bridge-building between North and South partners, public and private partners, and bringing in a range of diverse actors who would not normally meet at workshops to actually develop partnerships (not just talk about good ideas). Non-traditional for an MFA-organised conference is the use of popular practices mimicking TV; e.g. with an expert panel of devil's advocates evaluating new partnership proposals, or broadcasting storytellers from the future. The interviewees agree that the combination of meeting interesting people and actually getting down to work makes attendance worthwhile despite an overbooked calendar.

Conclusion

Why should we pay attention to environmental diplomacy in International Relations? And how should we seek to understand and explain environmental diplomacy? In general, new IR scholarship is challenging the hitherto weak position of diplomacy and diplomatic studies in IR. We see that the classic image of diplomacy has been replaced by an image of hybrid diplomacy with multiple actors, multiple issues and multiple practices. There are good reasons to highlight environmental diplomacy as emblematic, since this area of global governance will have a strong impact on the reorganisation of the global political economy facing climate change and other environmental threats. The orchestrating role that diplomacy plays in these new, overlapping environmental governance fields is important to study and understand if we want to be able to explain how global politics unfolds. Second, as argued in this chapter and in the international practices literature, international practices and narratives are central to better understanding and explaining the important role of hybrid diplomacy in global environmental governance.

What the cases of the Cartagena Dialogue in UNFCCC negotiations and the Global Green Growth Forum (3GF) and other green growth networks illustrate is that (new) communities in which the representatives of polities meet can be studied according to their practices, but more, we begin to notice how environmental diplomacy is shaped by and shaping the existing neoliberal world order. It is worth noting, too, that many networks assume the form of partnerships, which suggests that we should also focus on *partnerships* in practice as a central element in new narratives of global environmental governance.

Postscript following COP-21

This chapter was written prior to COP-21 in Paris, 30 November to 12 December 2015. The truly global Paris Agreement between 196 states warrants a postscript. A main conclusion by participants and observers is that French diplomacy made the agreement possible.

> Even as delegates celebrated at the conference's end, there was a palpable sense of relief from the exhausted French hosts. At many points in this fortnight of marathon negotiating sessions, it looked as if a deal might be beyond reach. That it ended in success was a tribute in part to their diligence and efficiency and the efforts of the UN. (Harvey, 2015)

Other observations fit the above conclusion. First, new partnerships as part of the diplomatic practices were instrumental. The US-China climate partnership launched in 2014 helped solve longstanding disagreements between these countries through e.g. a new 'enhanced policy dialogue' (NDRC, 2014: 8). And, in the last week of COP-21, a new partnership known as 'The High Ambition Coalition' was launched by a group of parties crossing the North–South divide. There is still some confusion as to establishment, participation and purposes; however, it became clear in the last two days of negotiations that the diplomatic efforts of leading countries and groups in this High Ambition Coalition (the Marshall Islands, Colombia, Mexico, the Gambia, the EU, the US, Norway, Canada, LDCs and AILAC) embodied the last push for the Paris Agreement. Second, central green growth actors were present at COP-21, and I would like to highlight GGGI's high-level event and launch of the Inclusive Green Growth Partnership which – because of its regional and financial anchoring – is likely to play a key role in the implementation of the Paris Agreement. The partners are the Asian Development Bank, the African Development Bank, the Inter-American Development Bank, and United Nations Regional Economic and Social Commissions – the Economic and Social Commission for Asia and the Pacific, the Economic Commission for Africa, the Economic Commission for Latin America and the Caribbean, and the Economic and Social Commission for Western Asia (GGGI, 2015).

*This chapter largely builds on my PhD dissertation (Blaxekjær, 2015) and engagement with the global environmental governance field.

References

3GF (2015). *3GF Partnerships*. Global Green Growth Forum. Retrieved from http://3gf.dk/en/about-3gf/3gf-partnerships/

Adler, E. (2008). The Spread of Security Communities: Communities of Practice, Self-Restraint, and NATO's Post—Cold War Transformation. *European Journal of International Relations*, 14(2), 195-230.

Adler E. and Pouliot V. (eds) (2011). *International Practices*. Cambridge: Cambridge University Press.

Adler-Nissen, R. (ed.) (2013). *Bourdieu in International Relations: Rethinking Key Concepts in IR*. London and New York: Routledge.

Adler-Nissen, R. (2015). Conclusion: Relationalism or Why Diplomats Find International Relations Theory Strange. In: O. J. Sending, V. Pouliot, and I. B. Neumann (eds). *Diplomacy and the Making of World Politics* (pp. 284-308). Cambridge: Cambridge University Press.

Adler-Nissen, R. and Pouliot, V. (2014). Power in Practice: Negotiating the International Intervention in Libya. *European Journal of International Relations, 20*(4), 889-911.

Audet, R. (2013). Climate Justice and Bargaining Coalitions: A Discourse Analysis. *International Environmental Agreements*, 13(3), 369-386.

Bauer, S. (2013). Strengthening the United Nations. In: R. Falkner (ed.). *The Handbook of Global Climate and Environment Policy, 1st Edition* (pp. 320-338). Chichester: Wiley.

Benedick, R. E. (1998). Ozone Diplomacy: *New Directions in Safeguarding the Planet*. Enlarged Edition. Cambridge, MA: Harvard University Press.

Bernstein, S. (2001). *The Compromise of Liberal Environmentalism*. New York: Columbia University Press.

Bigo, D. (2005). Global (In)security: The Field of the Professionals of Unease Management and the Ban-Opticon. In: N. Sakai and J. Solomon (eds). *Traces: A Multilingual Series of Cultural Theory*, 4 (Translation, Biopolitics, Colonial Difference) (pp. 34-87). Hong Kong: Hong Kong University Press.

Bigo, D. and Madsen, M. R. (eds) (2011). Bourdieu and the International – Special Issue. *International Political Sociology*, 5(3): 219-347.

Blaxekjær, L. Ø. (2015). *Transscalar Governance of Climate Change: An Engaged Scholarship Approach*. Copenhagen: Department of Political Science, University of Copenhagen. PhD series 2015/8.

Blaxekjær, L. Ø. and Nielsen, T. D. (2014). Mapping the Narrative Positions of New Political Groups under the UNFCCC. *Climate Policy*. Published online 17 October 2014.

Brown, C. (2012). The 'Practice Turn', Phronesis and Classical Realism: Towards a Phronetic International Political Theory? *Millennium: Journal of International Studies*, 40(3), 439-456.

Bull, H. and Watson, A. (eds) (1984). *The Expansion of International Society*. Oxford: Clarendon Press.

Bulkeley, H., Andonova, L. B., Betsill, M. M., Compagnon, D., Hale, T., Hoffmann, M. J., Newell, P., Paterson, M., Roger, C. and Vandeveer, S. D. (2014). *Transnational Climate Change Governance*. New York: Cambridge University Press.

C2ES (2015). *Toward 2015: An International Climate Dialogue. The Centre for Climate and Energy Solutions*. Retrieved from http://www.c2es.org/international/toward-2015

Casey-Lefkowitz, S. (2010, December 11). New International Agreement to Fight Climate Change Found Spirit for Consensus in Cartagena Dialogue Countries. S*witchboard: Natural Resources Defense Council Staff Blog*. Retrieved from http://switchboard.nrdc.org/blogs/sclefkowitz/new_international_agreement_to.html

Chasek, P. S. (2001). *Earth Negotiations: Analyzing Thirty Years of Environmental Diplomacy*. Tokyo: United Nations University Press.

Christoff, P. and Eckersley, R. (2013). *Globalization of the Environment. Plymouth*: Rowman & Littlefield.

Cooper, A. F., Heine, J. and Thakur, R. (eds) (2013). *The Oxford Handbook of Modern Diplomacy*. Oxford: Oxford University Press.

Dimitrov, R. S. (2010). Inside Copenhagen: The State of Climate Governance. *Global Environmental Politics*, 10(2), 18-24.

Dimitrov, R. S. (2013). International Negotiations. In: R. Falkner (ed.). *The Handbook of Global Climate and Environment Policy*, 1st Edition (pp. 339-357). Chichester: Wiley.

Eckersley, R. (2012). Moving Forward in the Climate Negotiations: Multilateralism or Minilateralism? *Global Environmental Politics*, 12(2), 24-42.

Elliot, L. (2013). Climate Diplomacy. In: Cooper, A. F., Heine, J. and Thakur, R. (eds) (2013). *The Oxford Handbook of Modern Diplomacy* (pp. 840-856). Oxford: Oxford University Press.

European Commission (2015, December 8). *EU and 79 African, Caribbean and Pacific Countries Join Forces for Ambitious Global Climate Deal*. Press release. Retrieved from http://europa.eu/rapid/press-release_IP-15-6273_en.htm?locale=en

Farrell, H. (2009). Institutions and Midlevel Explanations of Trust. In: K. S. Cook, M. Levi, and R. Hardin (eds). *Whom Can We Trust? How Groups, Networks, and Institutions Make Trust Possible* (pp. 127-148). New York: Russell Sage Foundation.

GGGI (2015, December 7). *New Global Initiative Launches at COP21 to Boost Green Growth Financing.* Global Green Growth Institute. Retrieved from http://gggi.org/new-global-initiative-launches-at-cop21-to-boost-green-growth-financing/

Harvey, F. (2015, December 14). Paris Climate Change Agreement: The World's Greatest Diplomatic Success. *The Guardian*. Retrieved from http://www.theguardian.com/environment/2015/dec/13/paris-climate-deal-cop-diplomacy-developing-united-nations

Hoffmann, M. J. (2011). *Climate Governance at the Crossroads: Experimenting with a Global Response after Kyoto. New York*: Oxford University Press.

Holmes, M. (2013). The Force of Face-to-Face Diplomacy: Mirror Neurons and the Problem of Intentions. *International Organization*, 67(4), 829-861.

Kaufmann, J. (1988). *Conference Diplomacy. An Introductory Analysis*. 2nd Revised Edition. Dordrecht: Martinus Nijhoff Publishers.

Lynas, M. (2011, March 10). *Thirty 'Cartagena Dialogue' Countries Work to Bridge Kyoto Gap*. Retrieved from http://www.marklynas.org/2011/03/thirty-cartagena-dialogue-countries-work-to-bridge-kyoto-gap/

Meilstrup, P. (2010). The Runaway Summit: The Background Story of the Danish Presidency of COP15, the UN Climate Change Conference. In: N. Hvidt and H. Mouritzen (eds). *Danish Foreign Policy Yearbook 2010* (pp. 113-135). Copenhagen: Danish Institute for International Studies.

Monheim, K. (2015). *How Effective Negotiation Management Promotes Multilateral Cooperation: The Power of Process in Climate, Trade, and Biosafety Negotiations*. New York: Routledge.

NDRC (2014). *Report of the U.S.-China Climate Change Working Group to the 6th Round of the Strategic and Economic Dialogue*, July 9, 2014. National Development and Reform Commission. Retrieved from http://en.ndrc.gov.cn/newsrelease/201407/P020140710297942139135.pdf

Neumann, I. B. (2002). Returning Practice to the Linguistic Turn: The Case of Diplomacy. *Millennium: Journal of International Studies*, 31(3), 627-651.

Neumann, I. B. and Leira, H. (eds) (2013). *International Diplomacy*. Four vols. London: Sage Publications.

Okereke, C. and Bulkeley, H. (2007). *Conceptualizing Climate Change Governance Beyond the International Regime: A Review of Four Theoretical Approaches.* Norwich: Tyndall Centre for Climate Change Research. Working Paper 112. Retrieved from http://www.tyndall.ac.uk/sites/default/files/wp112.pdf

Pouliot, V. (2008). The Logic of Practicality: A Theory of Practice of Security Communities. *International Organization*, 62(2), 257-288.

Pouliot, V. (2013). Methodology. In: R. Adler-Nissen (ed.). *Bourdieu in International Relations: Rethinking Key Concepts in IR* (pp. 45-58). New York: Routledge.

Pouliot, V. and Cornut, J. (2015). Practice Theory and the Study of Diplomacy: A Research Agenda. *Cooperation and Conflict*, 50(3), 297-315.

Randall, T. (2015, April 14). Fossil Fuels Just Lost the Race Against Renewables. *Bloomberg Business*. Retrieved from http://www.bloomberg.com/news/articles/2015-04-14/fossil-fuels-just-lost-the-race-against-renewables

Sending, O. J., Pouliot, V. and Neumann, I. B. (eds) (2015). *Diplomacy and the Making of World Politics*. Cambridge: Cambridge University Press.

Susskind, L. E., Ali, S. H. and Hamid, Z. A. (2014). *Environmental Diplomacy: Negotiating More Effective Global Agreements.* Oxford: Oxford University Press.

The New Climate Economy (2015). *Aims and Rationale.* Retrieved from http://newclimateeconomy.net/content/aims-and-rationale

UNFCCC (2014a). *Now, Up To and Beyond 2012: The Bali Road Map.* Retrieved from http://unfccc.int/key_steps/bali_road_map/items/6072.php

UNFCCC (2014b). *Durban: Towards Full Implementation of the UN Climate Change Convention*. Retrieved from http://unfccc.int/key_steps/durban_outcomes/items/6825.php

Wagenaar, H. (2011). *Meaning in Action. Interpretation and Dialogue in Policy Analysis*. Armonk and London: M. E. Sharpe.

Walker, J. and Ostrom. E. (2009). Trust and Reciprocity as Foundations for Cooperation. In: K. S. Cook, M. Levi, and R. Hardin (eds). *Whom Can We Trust? How Groups, Networks, and Institutions Make Trust Possible* (pp. 91-124). New York: Russell Sage Foundation.

Watson, A. (1982). *Diplomacy: The Dialogue between States*. London: Eyre Methuen.

Wenger, E., McDermott, R. and Snyder, W. M. (2002). *Cultivating Communities of Practice: A Guide to Managing Knowledge: Cultivating Communities of Practice*. Cambridge, MA: Harvard Business School Press.

Willis, A. (2010, November 29). UN Method Hangs in the Balance as Climate Talks Begin. *EU Observer*. Retrieved from http://euobserver.com/environment/31370

Wiseman, G. (2011). Bringing Diplomacy Back In: Time for Theory to Catch Up with Practice. *International Studies Review*, 13, 710-713.

Wiseman, G. (2015). Diplomatic Practices at the United Nations. *Cooperation and Conflict*, 50(3), 316-333.

Zelli, F. and Van Asselt, H. (eds) (2013). Introduction: The Institutional Fragmentation of Global Environmental Governance: Causes, Consequences, and Responses. *Global Environmental Politics*, 13(3), 1-13.

11

Climate Change, Geopolitics, and Arctic Futures

DUNCAN DEPLEDGE

ROYAL HOLLOWAY, UNIVERSITY OF LONDON, UK

Over the past decade, the Arctic has become the site of intense geopolitical intrigue among both practitioners and spectators of geopolitics and international relations (Borgerson, 2008; Ebinger and Zambetakis, 2009; Emmerson, 2010; Sale and Potapov, 2010). While the Arctic (or perhaps more accurately, the Arctic regions) has its own history (see McCannon, 2012), contemporary fascination with the northern latitudes is intimately linked with experiences and predictions of Arctic climate change. Although such experiences may be mediated in very different ways – from the indigenous witness of a changing landscape to the satellite images of sea-decline over time (see National Snow and Ice Data Centre, 2015) and the far-away analyst relying on forms of 'geopolitical remote sensing' (Moisio and Harle, 2006; Nuttall, 2012) – what they share is an understanding that the future Arctic is likely to have very little in common with the Arctic of the past. The Arctic is very much a region 'in change', and climatic changes are among the main drivers (ACIA, 2005; Koivurova, 2010).

Of course, speculation about what possible futures lie in store for the Arctic – especially if the Arctic becomes increasingly connected to 'progressive' global economic and social forces – is not new. In the early 1800s, the whaler William Scoresby brought news to the British Admiralty that the sea ice around Greenland was in retreat, paving the way for another attempt to traverse the Northwest Passage (NWP) by ship. In the early 1900s, Viljalmur Stefansson claimed the Arctic would become the next great hub of human and economic activity, and the only question remaining was who would be the main beneficiaries (Stefansson, 1921). In the 1930s, Joseph Stalin sought to conquer nature in the Russian Arctic to provide an economic resource base for the Soviet Union (Emmerson, 2010). And in the 1950s, the Canadian prime minister, John G. Diefenbaker (1958) presented his 'northern vision' of opening up Canada's Arctic frontiers to economic development. What many of these visions had in common was their failure to materialise.

In the 21st century, Arctic visionaries continue to abound, as do sceptics who point to past failures. What is different today is how closely their arguments are coupled with the 'reality' of the climatic changes in the Arctic regions. Their confidence in the Arctic's potential can most easily be assessed in the month of September each year, when the summer sea-ice minimum is reported, marking the end of the melt season (March–September). In recent years, new records have been set in 2002, 2007 and 2012 giving confidence to those predicting a significant increase of human activity in the Arctic. However, although sea-ice minima have not recovered to the average level recorded between 1979 (when satellite measurements began) and 2002, its continued annual variability has been seized upon by sceptics of an 'Arctic bonanza' as evidence that, overall, the Arctic environment remains hostile to increased human activities.

Consequently, observed and predicted climatic changes are increasingly important to the way in which both residents of the Arctic (living in indigenous communities, towns and cities local to the Arctic regions of the United States, Canada, Finland, Sweden, Norway, Iceland, Denmark (Greenland) and Russia) and stakeholders from beyond the Arctic think about what the future holds for the Arctic, especially in terms of whether the Arctic regions are 'opening up' (Grímsson, 2015) or need to be 'saved' (Greenpeace, 2015) from increased human activity. The result is that decisions are being taken at all levels of governance, from indigenous/local communities to global institutions about what kinds of activity to allow, how they should be pursued and, more broadly, how to situate the Arctic in relation to the wider world (e.g. as a new resource frontier, a shipping highway or a global commons).

The rest of this chapter looks at some of the key dimensions of this struggle over the Arctic's future, exploring how climate change has the potential to influence existing and future human activity in the Arctic and investigating the intersection of indigenous/local, national and international interests (and the various alliances this has produced) that is emerging as a result.

Climate change and human activity in the Arctic

Shipping

In the 15th century, European explorers sought navigable maritime passages through the Arctic region. Three possible routes were identified: a Northwest Passage (NWP) between the northern archipelagos of the North American continent; a Northeast Passage (NEP) following the northern coastline of the Eurasian landmass; and a Trans-Polar Route (TPR) straight across the Arctic Ocean. It is important to remember that at the time of the first European expeditions to seek out these passages, the Arctic region still represented a great unknown. Consequently, the search for northern passages from the 15th to the 19th century took on mythical, sublime, divine and even rational qualities as subsequent explorers sought funding for further expeditions to achieve a seemingly impossible transit through the Arctic (Spufford, 1996; Craciun, 2010).

While attempts on the TPR and NWP were all but abandoned by the 20th century, the challenge of conquering the NEP became a priority for the Soviet Union. Soviet planners believed that by developing the NEP, or Northern Sea Route (NSR), as the Russian portion of the route is known, the Soviet Union would possess the shortest route between the North Atlantic and the North Pacific, a significant strategic advantage over both its European and Asian rivals. Soviet domination of the route would also enable the unfettered transfer of economics resources across its huge swathes of territory (Laruelle, 2014). The NSR was subsequently closed to international traffic.

Traffic along the NSR peaked in 1987 before going into decline. In 1988, the Soviet Union began working with Norway and Japan on the International Northern Sea Route Programme to assess the economic feasibility of opening the NSR to international shipping (Dunlap, 1996). The programme established distance savings of 61 per cent between Hamburg and Dutch Harbour, Alaska; and 36 per cent between Hamburg and Yokohama. However, the waters of the NSR were also found to be relatively shallow, forcing larger vessels to travel more northerly routes, involving longer distances and more severe ice conditions. Today, the *Arcticmax*-classed container ships are at least three times smaller than those classed *Suezmax*, suggesting that the economic potential of using the route for trade remains small compared to the traditional trade routes via the Suez Canal (Humpert, 2013). Other significant factors reducing the economic potential of the NSR is the lack of ports along the NSR and the risk that changing ice conditions could lead to delays in an industry that increasingly relies on 'just-in-time' delivery (Humpert 2013).

In contrast, the NWP is far less developed. Although there were a number of expeditions during the 20th century to traverse the NWP with the support of icebreakers, it never became a commercially viable option. Ice conditions in the NWP are even harsher than in the NSR, a consequence of the fact that sea ice tends to drift across the Arctic Ocean towards North America, increasing the density of the ice there. Meanwhile, the TSR has remained blocked by the presence of all-year-round sea ice.

As this brief overview suggests, the technology to transit the NSR and the NWP has existed for decades. The challenges for would-be-users of these routes are therefore rooted in economics and risk. Businesses are unlikely to use these routes unless they produce significant cost savings. For much of the 20th century and early 21st century, any savings from reduced fuel costs associated with distance savings have been offset by the costs associated with building ice-strengthened hulls, the chartering of icebreaker support, skilling crews, the lack of markets, the risk of delays from changing ice conditions, and high insurance premiums (AMSA, 2009).

A key question being asked today is whether climate change has the potential to change this picture. Climate modellers suggest that the average sea-minimum is likely to continue falling over the coming decades (with some arguing it will completely disappear in the summer months). This has at least two material consequences: an increased area of 'open water' (albeit with a risk of icebergs) in the summer months; and a reduced area of 'multi-year' ice (AMSA, 2009). The possible gains are particularly obvious along the NSR where an increased area of open water or reduced ice thickness could allow transiting ships to travel further out from the coast, in deeper waters (allowing bigger ships to be used). A TPR could also open up even further north of the NEP. Even along the NWP, where Arctic sea ice becomes more concentrated, the waterways are becoming more navigable.

These observed changes have encouraged shipping companies (state-owned as well as a private) to look again at the viability of developing northern maritime passages. In 2009, for the first time two international commercial cargo vessels used the NEP to travel between Europe and Asia. In 2010 this increased to ten ships; in 2011 the number rose again to 34; in 2012 the number was 46, and in 2013 there were 71 commercial transits. The main benefits of using the Arctic routes are the distance savings that have the potential to reduce fuel consumption and greenhouse gas emissions. However, in 2014 the numbers fell back down to 23 ships after a year of much more severe ice conditions. Over the same period, destinational shipping to the Arctic (to resupply communities, for tourism, for fishing or to evacuate resources from oil and gas platforms and mines) has also increased dramatically, and while many remain sceptical of the potential for regular transits through the Arctic, further increases in destinational shipping seem likely, especially if the tourism, fishing and resource sectors boom and open waters become more prevalent for longer periods.

Oil and gas resources

Like shipping, the hunt for Arctic resources is not particularly novel. When European explorers set sail in search of the northern passages, they returned with new maps and reports documenting their encounters in Arctic waters. Their search for the northern passages was also supported by land expeditions across North America and Northern Russia. And while the maritime passages eluded them, what they did find was an abundance of living resources that would fuel the development of whaling, sealing, fur-trapping and fishing industries in Arctic lands and waters, almost to the point of extinction for the most lucrative species (Emmerson, 2010).

Growing global demand for crude oil and gas in the 20th century gave the Arctic a new material value. With growing enmity between Western world, the Middle East oil giants, and the Soviet Union, the Arctic regions were caught up in the search for new oil and gas fields. Less constrained by economic factors, the Soviet Union started to develop on-shore oil and gas fields in the Arctic in the 1930s. In North America, small amounts of oil were being pumped in Canada in the 1920s and 1930s, but it was not until the 1950s that the potential of Alaska was realised by the US Geological Survey, and more than a decade passed before the first major, commercially viable, discovery was made on-shore in Prudhoe Bay.

However, as in the case of Arctic shipping, the cost of developing oil and gas fields is far higher than it is in other parts of the world (such as the Middle East, Latin America and Africa). Even the use of complex technology for extracting oil and gas from shale reserves has proved more commercially viable. In part, this is because of the cost of operating in the Arctic, where conditions are especially challenging due to the cold weather extremes. More important though is the vast distances and the relative lack of infrastructure necessary to evacuate oil and gas to markets. For example, Prudhoe Bay's commercial viability is critically linked to the Trans-Alaska Pipeline, built in the aftermath of the 1973 oil crisis, when oil prices were high. When the price of oil falls, so does investment in the infrastructure needed to get it to market.

Today, the prospects of an Arctic oil and gas bonanza remain difficult to assess. Commercial interest in the Arctic region has increased significantly since the turn of the 21st century contributing to the popular idea that the Arctic is 'opening up' to human activity. A number of widely cited reports published by the US Geological Survey (USGS) estimated that the Arctic probably contained up to 13 per cent of the world's undiscovered oil, 30 per cent of its undiscovered natural gas and 20 per cent of its undiscovered natural gas liquids (USGS, 2015). However, these figures only estimated the quantities of fossil fuels 'technically recoverable' from the Arctic, without commenting on their commercial viability. In the first decade of the 21st century, the Organisation of the Petroleum Exporting Countries (OPEC) (2015) basket price of oil rose significantly, from US$24 a barrel in 2002 to US$94 a barrel in 2008. Despite a dip in 2009, between 2011 and 2013 the price rose again to over US$100 a barrel before crashing again in 2014 in what many analysts predict will be a sustained period of lower oil prices (around US$50 a barrel). A general rule of thumb for new developments in the Arctic is that their commercial viability rests on a basket price far closer to US$100 per barrel, meaning that the prospect of major oil and gas development in the Arctic has once again fallen for the time being. Nevertheless, this has not stopped a number of exploratory attempts by major oil companies, such as Shell, to locate proven reserves to add to their books which even if left undeveloped will buoy the stock price of their businesses.

Whether oil and gas fields in the Arctic regions are developed is for the most part a commercial question related to the global price of oil. Nevertheless, climatic changes could affect conditions in different ways. On the one hand, the retreat of summer sea ice creates a wider area of open water where the risk of ice to drilling infrastructure and support vessels is much reduced, meaning that it should prove easier to develop more offshore fields, especially in the Russian Arctic and the Beaufort Sea. Greater access along the NSR (see above) could also make it easier to ship resources recovered from Norway and Russia to markets in Asia where demand is currently greatest. On the other hand, the reduction of sea-ice cover and warmer atmospheric temperature increases the risk of storms and increased wave height (exacerbated further by global sea level rise), which could make drilling more difficult. Another factor to take into account is that as the Arctic warms, the permafrost layer that covers large parts of northern Russia, especially, is melting. The destabilisation of this permafrost layer is causing subsidence which in turn could disrupt the operations of on-shore oil and gas fields, as well as supporting on-shore infrastructure for off-shore developments. Such factors could increase the cost of some oil and gas operations in the Arctic, further chaining the commercial viability of many Arctic oil and gas development to the global price of oil (or heavy state subsidies, as seen, for example in Russia).

Climatic changes are also driving global interest in de-carbonisation, as represented by the two decades of international negotiations through the UN Framework Convention on Climate Change (UNFCCC). While many remain pessimistic that a global deal will ever be realised, there is widespread recognition among policymakers, businesses and scientists that the world will have to de-carbonise its economies if environmental catastrophe is to be averted (World Bank, 2015). The problem has been thrown into sharp relief by the Carbon Tracker Initiative (2015) which reported that to remain within 'safe' atmospheric carbon dioxide limits (calculated in terms of parts per million), the world as a whole cannot afford to burn all its existing fossil fuels reserves, let alone those which are unproven (as is the case for many of those estimated to exist in the Arctic). This has led a number of scientists and environmental campaigners to argue there is no point in trying to find and develop Arctic oil and gas reserves (McGlade and Ekins, 2015). Perhaps more significant, however, is that regardless of whether global de-carbonisation is achieved through a reduction in burning fossil fuels or an increase in the use of renewable energy sources (or a combination of both), the pressure on the basket price of oil will be the same; it will most likely fall, making Arctic development unviable from a commercial perspective. Consequently, the claim that climate change will open the door to an Arctic oil and gas bonanza, as in the case of *transit* shipping, remains heavily contestable.

Environment

The Arctic is often described by environmental campaigners as a pristine environment, rendered so by its long history of isolation from the industrial activity of humans. However, the Arctic Monitoring Assessment Programme (AMAP), a working group of leading international environmental scientists, are more cautious in their claims; arguing that while the Arctic may be considered 'one of the least polluted areas of wilderness on the planet', it is far from pristine (AMAP, 2015: 2).

Specifically, AMAP scientists point to the 'unique geographical, climatic and biological characteristics' (including prevailing atmospheric and oceanic currents, as well as large populations of mega-fauna such as whales and seals) that render the Arctic 'a "sink" for certain pollutants transported into the region from distant sources', including persistent organic pollutants (POPs) (such as a number of flame retardants and pesticides), heavy metals (e.g. mercury and lead) and radioactivity (in the form of radionuclides) (AMAP, 2015: 2). Further types of long-range pollution found in the Arctic include 'black carbon' – a form of soot that enters the atmosphere from the incomplete combustion of fossil fuels, biofuels and biomass –, as well as the discharging of oily wastes and the dumping of contaminated ballast water by ships (which may introduce invasive species into the Arctic ecosystem). Such pollutants can pose significant health risks to humans as well as the animals and plants of the wider Arctic ecosystem.

According to AMAP, climatic changes can interact with these pollutants in numerous ways (AMAP, 2011). For example, as the sea ice melts, previously immobilised contaminants including POPs, mercury and radionuclides may be taken up by the ecosystem (penetrating food chains). Similarly, contaminants trapped in the Arctic tundra (the largest sink for radioactive contaminants on Earth) are likely to be released into the surrounding environment as warmer temperatures drive permafrost melt. Overall, in a warmer Arctic, a whole range of contaminants are likely to become more mobile, spreading more widely across human communities and ecosystems.

Furthermore, if climatic changes do facilitate increased human activity in the Arctic, more localised forms of pollution are also likely to increase. Growing human populations and industrial activity will also produce more pollution from sewage flows, mining waste and the burning of fuels for heating, industrial processes and transportation. There is also an increased risk of pollution by oil spills whether through damage to infrastructure or support ships.

As a consequence, the (albeit contested) prospect that climatic changes could lead to increased human activity across the Arctic regions also brings with it a host of dilemmas about how best to protect the health of both local communities and the wider community. This has left many local leaders caught between on the one hand embracing new economic development opportunities and, on the other hand, trying to mitigate the increased risks to human health and the environment which threaten the viability of traditional ways of life (especially among the indigenous peoples of the Arctic). National leaders are similarly caught in a dilemma about how best to achieve sustainable economic development in the Arctic, while the transnational aspects of environmental pollution mean that the international community has also become embroiled in debates about how best to strike a balance between economic interests and environmental protection.

Intersecting interests

While the changing climate is not the only driver of broader changes in Arctic regions, it does matter to how various Arctic stakeholders are thinking about the Arctic's future. Changing perceptions, interests and activities relating to shipping, oil and gas resources and environmental pollution are brought together in claims that the Arctic is 'opening up' or 'needs saving'. Both claims are rooted in material changes, encompassing sea-ice melt and increased human activity. And both claims are producing new alliances among stakeholders at the local, national, regional and global level.

Broadly speaking, the claim that the Arctic is 'opening up' is supported by an alliance of indigenous peoples organisations, local leaders, scientists (modelling climate change and assessing environmental risks), international businesses (especially international oil companies [IOCs] but also others interested in fishing and mining), Arctic and non-Arctic nation-states, and regional organisations such as the Arctic Council. For example, in Alaska, Shell (an IOC) was working closely with the Alaskan government and the Arctic Slope Regional Corporation (ASRC) to search for oil fields in the Chukchi Sea. The decision to allow Shell to drill in the US Arctic region was supported by the federal government on the basis that it was important to national energy security. The Arctic Council established the Arctic Economic Council in 2014 precisely to promote such alliances between local/indigenous peoples, international businesses and national government. Countries such as the UK and Italy, which collect tax revenues from IOCs and seek to maintain stable global energy prices, add a further dimension of international support for the development of Arctic oil and gas fields. A similar story can be told for Greenland, Norway and Russia. In each case indigenous, local, national, international and economic interests, buttressed by scientific observations, models and assessments, mutually reinforce the view that climatic changes are creating new economic opportunities for a whole host of stakeholders.

At the same time, another alliance has emerged around the claim that the Arctic 'needs saving'. This alliance also involves indigenous peoples organisations, local leaders, scientists, international environmental non-governmental organisations, Arctic and non-Arctic nation-states and regional organisations such as the Arctic Council. In addition, the United Nations (UN) and the European Union (EU) are relevant players here due to their emphasis on establishing suitable regulatory frameworks to mitigate climate change, safeguard human activity (e.g. through rules on shipping) and protect the environment from pollution. Consequently, Shell's activities in Alaska have been contested by other indigenous peoples' organisations, such as the non-profit group called Resisting Environmental Destruction on Indigenous Lands (REDOIL), who have been working with international environmental NGOs (e.g. Greenpeace) to resist plans to drill for oil in the Chukchi Sea. Their claim that the Arctic is 'under threat' has a degree of global resonance to the extent that it is supported by civil society groups that want to 'save' the Arctic environment from both climate change and increased human activity.

While the nature of these alliances has perhaps been over-generalised here (there are, for example, also businesses and environmental NGOs which are working together in the Arctic), they are indicative of the way in which the geopolitics of the Arctic is being shaped by the intersecting interests and actions of a range of different stakeholders from both within and beyond the Arctic regions. This is a consequence of the Arctic's connectedness to the global environment, global economics, global technologies and global ethics. What happens in the Arctic does not stay in the Arctic. Likewise, what happens in the rest of the world does not stay out of the Arctic. Over the past decade, at least, observed climatic changes and predictions about future climate change (influencing issues relating to shipping, resource extraction and environmental pollution, among others) are affecting the ways in which these connections are thought about and pursued, especially in terms of whether the Arctic should be 'opened up' to increasing human activity or 'saved' from it.

Implications for governance

The tension between 'opening up' and 'saving' the Arctic is also putting pressure on regional and international governance structures. The viability of increased human activity in the Arctic regions will largely be determined by the regulatory frameworks, infrastructure and services (e.g. search and rescue) which are put in place (as well as those already existing such as the UN Convention on the Law of the Sea (UNCLOS) and the Arctic Council) to manage the exploration, extraction and evacuation of resources, in addition to other forms of commercial activity (mining, tourism and fishing, etc.). However, questions about what kinds of regulatory frameworks, infrastructure and services need to be put in place are difficult to divorce from questions about what kind of future Arctic climate should be anticipated. Decision-makers at local, national, regional and global levels are reliant on assessments of observed climatic changes, as well as climate models and projections about future climatic changes. Whether climate modellers predict an ice-free Arctic in 2016 or 2060 has enormous implications for decisions, for example, about the kinds of rules needed to be put in place for shipping activity and how quickly they are negotiated.

Further decisions about search and rescue services, infrastructure and environmental protection are similarly affected by what kinds of futures are imagined for the Arctic. For example, in the absence of global demand for Arctic oil and gas resources, or under pressure from global civil society, it may be the case that 'saving' the Arctic rather than 'opening it up' becomes the basis for future decisions about Arctic governance. However, it is also worth noting that, over the past decade, just as international interest in the commercial prospects of the Arctic has increased in those years when the sea ice appears to be in rapid retreat, so too has international interest in assessing the effectiveness of Arctic governance structures. Both experienced and anticipated climatic changes are therefore demonstrating their potential to affect the status quo of Arctic governance.

Conclusion

The visual spectacle provided by satellite images of retreating summertime sea ice in the Arctic makes it easy to assume that climatic changes are determining a new future for the Arctic by paving the way for increased human activity. However, decisions about whether to pursue different kinds of economic development in the Arctic are shaped by more than just environmental factors. Commercial and technical viability are key, and while climatic changes may lead to a reduction in sea ice, they also threaten to bring about more disruptive environmental conditions (such as a permafrost melt). Decisions about what kinds of governance arrangements should be put in place are not clear-cut either. These arrangements will be shaped by questions about what the future climate of the Arctic is expected to look like and whether this should lead to a greater focus on 'opening up' or 'saving' the Arctic regions. The struggle between these two possible futures (and there may be other futures to consider as well) will be fought by competing alliances that seek to mobilise shared interests and connections at all levels from the indigenous/local to the national, the regional and the international. Consequently, the issue of how to best anticipate and respond to climatic changes in the Arctic regions will be a problem not just of local or regional politics, but of global politics.

References

ACIA (2005). *Arctic Climate Impact Assessment*. Cambridge: Cambridge University Press.

AMAP (2011). *Combined Effects of Selected Pollutants and Climate Change in the Arctic Environment*. Retrieved from http://www.amap.no/documents/doc/combined-effects-of-selected-pollutants-and-climate-change-in-the-arctic-environment/747

AMAP (2015). *Summary for Policy-makers: Arctic Pollution Issues 2015*. Retrieved from http://www.amap.no/documents/doc/Summary-for-Policy-makers-Arctic-Pollution-Issues-2015/1195

AMSA (2009). *Arctic Marine Shipping Assessment*. Retrieved from http://www.pame.is/index.php/projects/arctic-marine-shipping/amsa

Borgerson, S. G. (2008). Arctic Meltdown: The Economic and Security Implications of Global Warming. *Foreign Affairs*, 87(2), 63-77.

Carbon Tracker Initiative (2015). *Carbon Tracker Initiative*. Retrieved from http://www.carbontracker.org/

Craciun, A. (2010). Frozen Ocean. *PMLA*, 125(3), 693-702.

Diefenbaker, J. G. (1958). *A New Vision*. Available at Arctic Sovereignty and International Relations. 3 August 2009. Retrieved from http://byers.typepad. com/arctic/2009/03/john-diefenbakers-northern-vision.html

Dunlap, W. A. (1996). Transit Passages in the Russian Arctic Straits. *International Boundaries Research Unit Maritime Briefings*, 7(1). Retrieved from https://www.dur.ac.uk/ibru/publications/view/?id=230

Ebinger, C. and Zambetakis, E. (2009). The Geopolitics of Arctic Melt. *International Affairs*, 85(6), 1215-1232.

Emmerson, C. (2010). *The Future History of the Arctic*. London: The Bodley Head.

Greenpeace (2015). *Save the Arctic*. Retrieved from https://www. savethearctic.org/

Grímsson, O. R. (2015). Iceland's President: 'Arctic Open for Business'. BBC World Service. *News Hour Extra*. Interview by Owen Bennett Jones. 20 October 2015. Retrieved from http://www.bbc.co.uk/programmes/p035qsl5

Humpert, M. (2013). T*he Future of Arctic Shipping: A New Silk Road for China?* Washington, D.C.: The Arctic Institute. Retrieved from http://issuu. com/thearcticinstitute/docs/the_future_of_arctic_shipping_-_a_n

Koivurova, T. (2010). Limits and Possibilities of the Arctic Council in a Rapidly Changing Scene of Arctic Governance. *Polar Record*, 46(237), 146-156.

Laruelle, M. (2014). *Russia's Arctic Strategies and the Future of the Far North*. London: M.E. Sharpe.

McCannon, J. (2012). A *History of the Arctic: Nature, Exploration and Exploitation*. London: Reaktion Books.

McGlade, C. and Ekins, P. (2015). The Geographical Distribution of Fossil Fuels Unused When Limiting Global Warming to 2°C. *Nature*, 517, 187-190.

Moisio, S. and Harle, V. (2006). The Limits of Geopolitical Remote Sensing. *Eurasian Geography and Economics*, 47(2), 204-210.

National Snow and Ice Data Centre (2015). *Arctic Sea Ice News & Analysis*. Retrieved from http://nsidc.org/arcticseaicenews/

Nuttall, M. (2012). Tipping Points and the Human World: Living with Change and Thinking about the Future. *AMBIO*, 41(1), 96-105.

OPEC (2015). *OPEC Basket Price*. Retrieved from http://www.opec.org/opec_web/en/data_graphs/40.htm

Sale, R. and Potapov, E. (2010). T*he Scramble for the Arctic: Ownership, Exploitation and Conflict in the Far North*. London: Frances Lincoln.

Spufford, F. (1996). *I May Be Some Time: Ice and the English Imagination*. London: Faber and Faber.

Stefansson, V. (1921). *The Friendly Arctic: The Story of Five Years in Polar Regions*. London: Macmillan.

USGS (2015). *Circum-Arctic Resource Appraisa*l. News & Recent Publications. Retrieved from http://energy.usgs.gov/regionalstudies/arctic.aspx

World Bank (2015). *Decarbonizing Development: Three Steps to a Zero-Carbon Future*. Washington, DC: IBRD/World Bank. Retrieved from http://www.worldbank.org/content/dam/Worldbank/document/Climate/dd/decarbonizing-development-report.pdf

12

Renewable Energy: Global Challenges

LADA V. KOCHTCHEEVA
NORTH CAROLINA STATE UNIVERSITY, USA

Developing renewable energy sources contributes to alleviating poverty, fuelling industrial production and transportation, expanding rural development and protecting health while promoting sustainability and environmental quality (Hostettler, 2015). Renewables account for approximately 20 per cent of global final energy consumption, with the most prominent growth happening in the power sector and with global capacity rising more than 8 per cent in 2013 (IEA, 2014a). Fossil fuels, however, continue to dominate global primary energy consumption, with coal remaining the major contributor to the world's energy pool (REN21, 2014). Almost 1.3 billion people in the world, mainly in rural areas, live without access to electricity and 2.7 billion without modern reliable energy services (UNDP, 2013; Alliance for Rural Electrification, 2014; IEA, 2014a). Global energy consumption is projected to rise by 56 per cent by 2040, with fossil fuels dominating the energy grid (US EIA, 2013). Strong economic and continued population growth in developing countries will be the prevalent force driving world energy markets during that period. Coal use is on the rise, mainly due to China's consumption, and global energy-related carbon dioxide emissions are predicted to have a 46 per cent increase by 2040, a rise from about 31 billion metric tons in 2010 (US EIA, 2013).

These developments have been prompting efforts to deploy renewable energy sources in many countries of the world to make access to energy more sustainable and address the problems of air quality and climate change. The United Nations (UN) has declared the years 2014–2024 the decade of Sustainable Energy for All (United Nations, 2015). Renewable energy technologies, which are a part of the low-carbon facet of global energy supply, are rapidly increasing their presence in many countries of the world. The top five countries for total installed renewable power capacity by the beginning of 2014 were China, the United States (US), Brazil, Canada and Germany. In the European Union (EU), renewables have represented the majority (72 per cent) of new electric generating capacity for the last several years (REN21, 2014). Renewables, however, are no longer dependent on a small number of countries. Major renewable energy companies became very interested in Africa, Asia and Latin America; where new markets are emerging on and off-grid. Investment patterns are also shifting away from traditional governmental and foreign donor sources to greater reliance on private and often local firms and banks (Martinot et al., 2002; REN21, 2014). Support for the adoption of renewable energy has been growing among the governmental agencies, industry, non-governmental organisations and the public at large. These actors pursue energy, environment and development agendas at local, regional and global levels (Bayer et al., 2013; REN21, 2014).

The policy, manufacturing and financing for renewables continue to expand across the developing world and emerging economies. By 2018, according to the International Energy Agency (IEA), non-OECD countries are predicted to account for 58 per cent of total renewable generation, up from 54 per cent in 2012. Renewable energy generation in most developing countries still mostly depends on inexpensive and abundant hydropower, but other technologies are on the rise in countries with good resources and emerging support measures (IEA, 2013).

Helped by global subsidies, renewables may account for almost half of the increase in total electricity generation to 2040, with the use of biofuels more than tripling (IEA, 2014b). Generation of renewables is also predicted to rise more than twice as much in many developing countries and emerging economies. The number of developing countries with policies in place to support renewable energy has increased sixfold since 2006, resulting in one-fifth of the world's power production presently coming from renewable energy sources (United Nations, 2014). Continuing advances in technology, innovations in policy and financing, decreasing prices, and educational efforts make renewables more attractive and affordable for a larger number range of consumers around the world (REN21, 2014).

However, as renewable energy policies, markets and industries develop, they increasingly face new challenges, which are multifaceted and highly complex. The fact that significant reserves of fossil fuels are still available impedes the willingness to give sufficient importance to the renewables. Fossil fuels receive six times more in subsidies than renewable energy sources (Hostettler, 2015). In their competition with mature fossil fuel and nuclear technologies, renewables encounter major challenges to commercialisation, including underdeveloped infrastructure and lack of economies of scale. Additionally, the integration and combination of different energy sources from a market, policy and technical perspectives are becoming more challenging and requiring capacity building. The success of deploying new technologies depends on the ability to build, monitor and maintain energy infrastructure, as well as train scientists, decision-makers and manufacturers at domestic and global levels (MacLeod and Rosei, 2015).

For developing countries, especially, costs and the lack of sound policies are some of the main barriers. Start-up expenditures, the lack of approaches to balance price disparities between renewables and fossil fuels, and overarching structural obstacles, such as the centralised nature of the energy industry, impede support and implementation of new initiatives, deter investment in renewables and frustrate more localised approaches to energy access. Notably, the introduction of renewables presents an issue of inequality. The problem is that the rate of technology diffusion, the availability of financing and policy implementation are uneven within and across countries' national boundaries. And while renewable energy is one of the world's fastest-growing energy sources now, increasing by 2.5 per cent per year (REN21, 2014), it has not been sufficient to keep pace with the consequences of rapid growth in demand for energy.

Employing renewable energy faces a range of economic, policy, structural and social challenges, requiring not only further technological development and investment but also a deeper understanding of both the success factors and the obstacles to accomplish widespread adoption. This chapter will proceed by presenting the discourse on the deployment of renewable energy with an emphasis on policy, technology and investment for renewables in the developing countries. It will continue with the discussion of some of the major international challenges that may explain the difficulties in the adoption and implementation of renewable energy, including the effects of global learning on the introduction of renewables, the barriers to technology and policy diffusion. The chapter ends with concluding remarks.

Policy, technology and investment considerations for renewable energy

Countries around the world increasingly take measures to research and deploy renewable energy sources to improve energy security, encourage economic growth and respond to environmental challenges particularly associated with climate change. The research by the International Energy Agency demonstrates that renewable energy technologies have been mainly adopted by countries with relatively high gross domestic product (GDP) per capita and also high energy security concern (Müller et al., 2011). Such front-runner countries have both the capacity and the impetus to engage with renewables especially during the initial development stages, when costs are high. The wealth of these countries also influences the choice of the technology for generating renewables, where countries with lower economic capacity focus on lower-cost, well understood and established renewable sources, such as hydro and biomass. With the increasing maturity of renewables, falling prices, enhanced education and improving competitiveness, the likelihood of technology diffusion across national boundaries increases. For many developing countries, the opportunities to deploy renewable energy sources exist particularly in cases where the resource conditions are good and the need for expansion in energy access is high (Müller et al., 2011).

The majority of developing countries are blessed with substantial renewable energy resources, such as solar and wind, covering large geographical areas and not requiring a centralised approach for dissemination. The deployment of renewable energy can make effective use of available human capital in countries with underemployment without compromising the desirable features of energy supply. The rationale for the adoption of renewables is strengthened both by the improvement in the quality of life of rural, distant, under-served populations and by the devastating environmental effects of fossil fuel use. Additionally, for developing countries the stakes of dealing with environmental consequences are much higher than for developed nations. Yet, the capacity of developing countries to manage severe environmental degradation and its health consequences is often inadequate and undermined by their vulnerability to external shocks either financial or environmental.

Developing countries also have a number of common characteristics that influence the acceptance, spread and sustainability of renewable energy approaches (Kandpal et al., 2003). The most important feature is the desire for economic development and a constant trade-off between growth and environmental protection (Mohiuddin, 2006; Hostettler, 2015). The majority of the population has low, or relatively – compared to the developed countries – lower energy consumption per capita, which reflects poorer quality of life and low purchasing power of potential end-users of renewable energy. Many of the developing countries import fossil fuels, which creates risks for energy security and foreign exchange. And while developing countries have experience with renewable energy technologies, projects, especially in the past, were characterised by fragmented efforts and were implemented in isolation from other development challenges such as health, education and local development. Additionally, up until the 1990s, renewables were introduced without the guidance of integrated policies (GNESD, 2007). So, the priority for many developing countries became the creation of supporting policies for the adoption of renewables.

Policy

The introduction and implementation of support policies to a large degree determine the extent to which renewables are developed in a country (Berg, 2013). The renewable energy market is also a policy-driven market. The adoption of support policies, however, does not follow the one-size-fit-all approach. The choices of policy instruments and sectors need to reflect the objectives of each country according to its priorities regarding environmental protection, economic development and socio-economic structure (Djiby, 2011). Also, while a particular policy approach may be considered as effective, public expenditures required to achieve this might be disproportionate and therefore politically unbearable. Determining the costs and risks of various policy tool kits involves multiple, complex assumptions and considerations of country market structure, resource endowments and national goals (UNDP, 2013).

Developed countries usually serve as front runners in establishing new policies. For example, in Europe, new policies are emerging to accelerate or manage the integration of renewables into existing power systems, including the development of energy storage and smart grid technologies. Developing countries are adopting support policies and experimenting with various policy tools. By the end of 2013, developing and emerging economies became the leaders in the increase of renewable energy support policies and accounted for 95 of the 138 countries with such policies (REN21, 2014). Renewable energy support policies usually include the use of regulatory and economic instruments such as standards, planning and codes; building institutional structures and capacity; as well as voluntary approaches, including information provision, advertisement, and education. The latter policy tools are only in their nascent stages in developing countries, with most emphasis on economic tools such as direct investments in infrastructure, fiscal and financial incentives and market-based initiatives, including allowances for greenhouse gas (GHG) emissions or green certificates. Specifically, feed-in tariff policies (policies based on prices) and renewable portfolio standards (RPS) (policies based on quantities) are the most commonly used policy support mechanisms. Regulatory policies, as well as economic instruments, have been found to have a strong effect on the production of renewable energy in the developing world. (Pfeiffer and Mulder, 2013)

Most renewable energy policies enacted or revised focus on the power sector, yet a big challenge for the renewable energy industry, in general, has been competition from heavily subsidised conventional energy. Other significant policy challenges for the developing countries and emerging economies include the problems of policy formation in the context of economic development, where growth is a priority and where old and entrenched mechanisms are difficult to part with. Households or energy companies which want to install wind turbines or solar panels have been discouraged by lengthy pay-back times. Without political measures to facilitate access to the market and increased consumer demand, manufacturers of wind turbines or solar photovoltaic (PV) panels cannot produce the unit volumes needed to bring prices down and drive technological innovation.

Another challenge is the creation of an enabling policy environment and targets that can encourage the private sector to participate in financing the development of renewable energy projects (UNDP, 2013). Supporting renewable energy demonstration projects to spread information in remote areas, training microfinance leaders and decentralising the implementation of renewable energy projects may foster the spread of renewable energy projects. Such actions would also help build a better equipped sustainable renewable power industry, generate profits and create jobs, as well as increase efficiency in financing (Mohiuddin, 2006; MacLeod and Rosei, 2015). Most support for renewable energy policies and technologies in developing countries comes from local governments or from international donors, which undermines their sustainability, as the funds fluctuate with changing priorities and crises.

Finally, the establishment of innovative policies and sustainability of renewable energy markets and technology may benefit from the adoption of an overall energy governance framework. The introduction of energy governance enables more efficient involvement of various stakeholders, increasing the decision-making authority of local governments, creating diversified institutional arrangements and public involvement (Djiby, 2011), as well as increasing capacity to tailor policy to local conditions, especially in countries with wealth disparities and varied commitments to environmental improvement.

Technology

The choice and deployment of renewable energy technologies may significantly contribute to building a comprehensive strategy towards more sustainable economic growth. Technological innovation and capacity in renewables result from a broad range of factors and not merely from effective research and development efforts (Müller et al., 2011). These include technological capability of a country, innovation-friendly regulation, market conditions that favour adaptive learning, and others. Specifically, a study of patent activity demonstrates the relative strength of different developed countries in generating technology innovation and using their pioneering country advantage in renewables. For instance, Germany and Denmark exhibit strength in wind energy technologies, the United States, Germany and Japan show the highest shares of patents for solar PV technology, and the EU as a whole presents the largest patent shares for biomass and biogas, wind and solar thermal technologies (Müller et al., 2011). The emerging challenge with the adoption and spread of renewable energy technologies is twofold. The first is whether these pioneer countries can sustain their first-mover advantage in the face of growing competition from emerging economies with lower production costs. While renewable energy innovation has traditionally been the prerogative of the developed world, it is now on the rise in the emerging economies. BRICS countries rank among the top global inventors (Bayer et al., 2013). The biggest limitation, however, is that these countries do not yet export their technologies to either developed or other developing countries on any substantial scale.

The second concern is whether many low-income developing countries are able to secure the diffusion of these technologies, as well as create conditions for the development of domestic renewable energy technologies (Ockwell and Mallett, 2012). Due to the relatively high upfront costs of most technologies, having access to finance is considered to be an important prerequisite for their adoption (Kandpal et al., 2003; Brunnschweiler, 2010; Huenteler et al., 2014). As such, higher level of economic development tends to influence the level of renewable energy development, because the former usually suggests more public and private financial resources, increasing environmental awareness and growing electricity demand (Pfeiffer and Mulder, 2013). Globally, there has been a discernible trend in lowering costs and improving efficiency of renewable technology installations, making it possible to build onshore wind and solar PV installations in certain areas around the world without subsidy support, particularly in Latin America. There has also been an increased use of mini-grids, which supported the dissemination of renewable energy-powered electrification in rural and suburban areas with poor electrification (Müller et al., 2011). With the help of information and communication technology for power management and end-user services, technical advances that allow the integration of renewable sources in mini-grid systems, stimulated a rapid expansion of the use of renewables-powered mini-grids in developing countries (REN21, 2014).

Investment

Renewable energy sources are progressively being viewed as investments that can generate economic advantages by reducing dependence on foreign fossil fuels, improving air quality and health safety, increasing energy access and security, building opportunities for economic development and reducing unemployment. Global investment in renewable power capacity and fuels increased more than fivefold over the period 2004–2013. Total global investment (both public and private) in research and development for renewable energy technologies has almost doubled over the past decade (REN21, 2014).

The portrait of renewable energy development is, however, becoming more multifaceted, with more challenges seen in some regions of the globe. While new global investment in renewable energy remains relatively high, there is observable decrease in the last several years. Global new investment in renewable power, without hydropower projects, was US$214.4 billion in 2013 (REN21, 2014), which was down 14 per cent relative to 2012, and 23 per cent lower than the record level in 2011 (Frankfurt School-UNEP, 2014). The reduction in investments for two consecutive years was mostly due to uncertainty over support policies in Europe and the United States and retroactive reductions in support in some other countries. While Europe's investment was down by more than 40 per cent, the emerging economies are coming to the forefront for the first time, with China alone having invested more in renewable energy than all the European countries (REN21, 2014). Economic difficulties, policy uncertainties, reductions in incentives and strong and persistent competition from traditional energy sources played the role in the investment volume. Different countries in the world specifically experienced challenges in integrating renewables in their power grids, while the manufacturing sector, especially wind and solar, moved into a complex phase of restructuring and consolidation (IEA, 2013).

Additionally, renewable energy sources are being introduced into an uneven playing field, where their energy prices do not fully reflect externalities. Global subsidies for traditional fuels and nuclear energy remain high despite the benefits of renewables and environmental quality concerns. Estimates for the global cost of fossil fuel subsidies range from $544 billion to $1.9 trillion – several times higher than those for renewable energy (REN21 2014). A large part of renewable capacity additions is found in countries with extensive subsidy systems, which can compensate investors for the comparatively high costs of the renewable energy technologies (Wagner, 2014). Although renewable energy technologies have undergone significant cost reductions in the last several years, they are still comparatively immature and much less able than traditional sources of energy to provide cost-competitive power generation on a large scale. Especially in developing countries, the barriers towards a larger transition to renewable energy are not just the disproportionate subsidies and technology costs but also the challenges of securing long-term commitment and affordable finance (UNDP, 2013).

Global learning and diffusion of policy and technology

Because high upfront costs and disproportionate financing remain some of the major challenges for large-scale commercialisation and adoption of renewables, especially in developing countries, the question becomes how to facilitate the development and diffusion of renewable energy technologies and policy approaches (Schmidt, 2014). As the industries producing renewable energy technologies are becoming increasingly globalised (Huenteler et al., 2014), the conditions of global learning, the paths of technology diffusion and the characteristics of the front-runner and borrower countries influence adoption and transfer of renewable energy technologies.

Global learning

The spread of renewable energy technologies, especially in developing countries and emerging economies, depends on the combination of global and local learning processes, which, in turn, depend on domestic and international policy provisions and local institutional and industrial contexts (Huenteler et al., 2014). Building technological capabilities through learning is viewed as an important contributing factor in the deployment of renewable energy sources that can result in cost reductions, performance improvement and climate change mitigation efforts (Ockwell and Mallett, 2012; Lema and Lema, 2013). Increased technological capacity — the accumulation of technological knowledge and experience — is essential for building local capacity for production, poverty reduction and socio-economic development. However, technological capabilities comprise not only the information, materials and components but also the skills and well developed routines. This means that technological learning demands the development of local capacities in addition to the removal of trade barriers, provision of intellectual rights and other forms of technical assistance (Huenteler et al., 2014).

Individuals and firms learn and innovate via their collaboration with research institutes, consumers, suppliers, competitors, etc. The formation of formal and informal networks, as well as a system of financing for research and development, is an important requisite for technological learning. Domestic positive policy and investment climate may influence and increase technological capacity through learning, yet it is not the single function for technology advancement. Acquired technological capacities can risk decline in the absence of a domestic support policy framework, stable financing and an accepting culture. The significant task is to create domestic opportunities and an atmosphere for continuous learning for governments, firms and communities through the build-up of organisational processes and culture, support of science and education, as well as various systems for innovation.

Technological learning has an ever-evolving global component composed of the movement of goods, services, materials, documents and information where many parts of supply chains are geographically dispersed and disintegrated. The markets for renewable energy technologies have also become globalised. The aggregate global market knowledge and trends to a large degree stipulate the development of technological capacities in firms and industries beyond domestic and local circumstances (Huenteler et al., 2014). Timely and reliable data on renewable energy are also crucial for creating energy plans, outlining criteria for targets, examining progress and effectiveness of policy actions and generating investment. Global data collection on renewables demonstrates an improvement with more broad and sensible record keeping, increased openness and better communication among stakeholders. However, there are many challenges remaining. In many countries, data on renewable energy are incomplete, not collected systematically, and with a time lag between developments and availability, which can be a significant impediment to relevant and timely decision-making process. Additionally, the large number and diversity of technologies in certain sectors, such as heating and cooking, may also lead to the dispersed and inconsistent data collection (REN21, 2014). This, in turn, can cripple the capacity to make informed decisions and affect financing opportunities, policy outcomes and planning for the future.

Diffusion of policy and technology

The dynamics of the global system and information flows also stand out as an important set of requisites influencing diffusion of renewable energy policy and technology. Diffusion is a process by which policy, technology and innovations are communicated throughout the international system over time (Jörgens, 2005). The role of international organisations, transnational networks and political linkages between states, the influential role of frontrunner countries and the institutionalisation of policy transfer shape the mechanisms of diffusion (Kern et al., 2001; Tews, 2005). Among many international factors, the role of front-runner states is critical for diffusion, as their expertise, economic strength, demand for environmental solutions and desire to influence others all impact on whether a policy diffuses effectively (Graham et al., 2013). Acting first in the adoption of renewable technology may give a country the ability to 'defend their own interests by assuming an active, pioneering role' (Kern et al., 2001: 5). In addition to regulatory advantages, leader nations may have a market advantage in renewable technology. Policy innovations in leader countries can set international standards, which put pressure on other countries to adopt similar policy (Kern et al., 2001). This regulatory conformism can lead to group behaviour in policy making and contribute to diffusion.

Diffusion of renewable energy policy and technology embodies the flows of information, experience and equipment for the adoption of renewable energy sources among various stakeholders, such as governments, firms, financial institutions and other entities. The diffusion may provide an adopting country the capacity to implement, operate and maintain borrowed technologies and policy measures to local conditions. The spread of renewable technology and policy may not be easy and straightforward for developing countries where immediate financial and institutional constraints are likely to be more acute than in most developed countries. Diffusion mechanisms must be responsive to the particular needs and challenges of developing countries and must advance, to the greatest extent feasible, multiple societal objectives. In countries where a significant portion of the population still lacks access to basic stable energy services, concerns about long-term environmental sustainability often are surpassed by more pressing problems of energy access and affordability.

In general, it has been demonstrated that the diffusion of renewable energy technologies depends on the implementation of economic and regulatory instruments, per capita income and schooling levels and stability of the regime (Pohl and Mulder, 2013). A thorough understanding of domestic resources and knowledge of policy measures that have been successfully applied in other countries also influences the prospects of adoption of efficient and context-specific measures for the renewable sources. A combination of measures to build a coherent enabling framework is important to ensure cost-efficient transfer and diffusion of a specific technology. Especially in developing countries, there is a need for targeted technical assistance that incorporates social equity components, not jeopardising low-income consumers but still attractive to both private and international aid organisations (UNEP, 2011).

Conclusions

Although the adoption of renewable energy sources is increasing in many parts of the world, widespread adoption is constrained by a multitude of policy, regulatory, technological, social and financial barriers. Enormous subsidies for fossil fuels and nuclear power persist, and they continue to vastly outweigh financial incentives for renewables. Market failures coupled with unfavourable financial, institutional and regulatory environments demand governmental intervention to establish renewable energy sources. Building human and institutional capacity, setting up research and development infrastructure, creating an enabling environment for investment and providing information present a challenge for many countries. A lack of supporting policy framework also stands as a large barrier among the risks that can undermine renewable project feasibility, even in the conditions of plentiful resources and favourable technology development. This array of challenges to the adoption of renewables requires a systematic approach in research to deepen the understanding of the challenges that exist for the deployment and diffusion of renewables in different countries. The exact difficulties that countries face depend upon national circumstances, the dynamics of the global system and the flows of information and resources. Devising effective responses to a problem that is global and multi-generational in scale presents a challenge that is, especially for developing countries, greatly complicated by the simultaneous need to expand access to essential energy services and to advance multiple objectives, including economic and social development goal as well as environmental ones.

References

Alliance for Rural Electrification (2014). *Energy Access in the World*. Available at: http://www.ruralelec.org/

Bayer, P., Dolan, L. and Urpelainen, J. (2013). Global Patterns of Renewable Energy Innovation. *Energy for Sustainable Development*. 17(3), 288-295.

Berg, S. (2013). Regulatory Functions Affecting Renewable Energy in Developing Countries. *The Electricity Journal*, 26(6), 28-38.

Brunnschweiler, C. N. (2010). Finance for Renewable Energy: An Empirical Analysis for Developing and Transition Economies. *Environmental Development Economics*, 15(3), 241-274.

Djiby, R. T. (2011). An Energy Pricing Scheme for the Diffusion of Decentralized Renewable Technology Investment in Developing Countries. *Energy Policy*, 39(7), 4284-4297.

Frankfurt School-UNEP (2014). FS-UNEP *Collaborating Center. Global Trends in Renewable Energy Investment 2014*. Key Findings. Frankfurt School of Finance and Management.

GNESD (2007). *Renewable Energy Technologies and Poverty Alleviation: Overcoming Barriers and Unlocking Potentials*. Global Network on Energy and Sustainable Development.

Graham, E., Shipan, R. and Volden, C. (2013). The Diffusion of Policy Diffusion Research in Political Science. *British Journal of Political Science*, 43, 673-701.

Hostettler, S. (2015). Energy Challenges in the Global South. In: S. Hostettler, A. Gadgil, and E. Hazboun (eds). *Sustainable Access to Energy in the Global South: Essential Technologies and Implementation Approaches* (pp. 3-9). London: Springer International Publishing.

Huenteler, J., Niebuhr, C. and Schmidt, T. S. (2014). The Effect of Local and Global Learning on the Cost of Renewable Energy in Developing Countries. *Journal of Cleaner Production.*

IEA (2013). *Renewable Energy. Medium Term Market Report 2013. Market Trends and Projections to 2018.* OECD/IEA.

IEA (2014a). *World Energy Outlook 2014.* OECD/IEA.

IEA (2014b). *Renewable Energy. Medium Term Market Report 2014.* Executive Summary. OECD/IEA.

Jörgens, H. (2005). Diffusion and Convergence of Environmental Policies in Europe. *Environmental Policy and Governance,* 15(2), 61-62.

Kandpal, T. C., Purohit, P., Kumar, A. and Chandrasekar, B. (2003). Study of Selected Issues Pertaining to the Economics of Renewable Energy Utilization in Developing Countries. *Journal of the Solar Energy Society of India,* 13(1-2), 57-82.

Kern, K., Jorgens, H. and Janicke, M. (2001). T*he Diffusion of Environmental Policy Innovations: A Contribution to the Globalisation of Environmental Policy.* WZB Working Paper No. FS II 01 – 302. Berlin: WZB.

Lema, A and Lema, R. (2013). Technology Transfer in the Clean Development Mechanism. *Global Environmental Change,* 1, 301-313.

MacLeod, J. M. and Rosei, F. (2015). Supporting the Development and Deployment of Sustainable Energy Technologies Through Targeted Scientific Training. In: S. Hostettler, A. Gadgil, and E. Hazboun (eds). *Sustainable Access to Energy in the Global South: Essential Technologies and Implementation Approaches* (pp. 231-233). London: Springer International Publishing.

Martinot, E. Chaurey, A., Lew, D., Moriero, J. R. and Wamukonya, N. (2002). Renewable Energy Markets in Developing Countries. *Annual Review of Energy and Environmen*t, 27, 309-348.

Mohiuddin, S. (2006). Microfinance: Expanding the Role of Microfinance in Promoting Renewable Energy Access in Developing Countries. *The Georgetown Public Policy Review*, 11, 119.

Müller, S., Brown, A. and Ölz, S. (2011). *Renewable Energy*. Policy Considerations for Deploying Renewables. Information Paper. OECD/IEA.

Ockwell, D. and Mallet, A. (2012). *Low-Carbon Technology Transfer: From Rhetoric to Reality*. New York: Routledge.

Pfeiffer, B. and Mulder, P. (2013). Explaining the Diffusion of Renewable Energy Technology in Developing Countries. *Energy Economics*, 40, 285-296.

Pohl, B. and Mulder, P. (2013). *Explaining the Diffusion of Renewable Energy Technology in Developing Countries*. German Institute of Global and Area Studies. Paper No. 207. Hamburg: GIGAS.

REN21 (2014). *Renewables 2014*. Global Status Report. Retrieved from: http://www.ren21.net/Portals/0/documents/Resources/GSR/2014/GSR2014_full%20report_low%20res.pdf

Schmidt, T. S. (2014). Low-carbon Investment Risks and De-risking. *Nature Climate Change*, 4, 237-39.

Tews, K. (2005). The diffusion of environmental policy innovations: Cornerstones of an analytical framework. European Environment: *The Journal of European Environmental Policy*, 15(2), 63-79.

UNDP (2013). *Derisking Renewable Energy Investment*. New York: UNDP.

UNEP (2011). *Diffusion of Renewable Energy Technologies: Case Studies of Enabling Frameworks in Developing Countries*. Technology transfer Perspective series.

United Nations (2014). *Developing Nations' Policies Push Renewable Energy Capacity to Record High, Says UN-backed Report*. Retrieved from http://www.un.org/apps/news/story.asp?NewsID=47952#.VKxDvyvF_7c

United Nations (2015). *United Nations Decade of Sustainable Energy for All*. Retrieved from http://www.se4all.org/decade/

US EIA (2013). *International Energy Outlook 2013*. DOE/EIA–0484.

Wagner, F. (2014). *Renewables in the Future Power Systems: Implications of Technological Learning and Uncertainty*. New York, Dordrecht, London: Springer.

13

The Fossil Fuel Divestment Movement within Universities

LEEHI YONA

DARTMOUTH COLLEGE, USA

&

ALEX LENFERNA

UNIVERSITY OF WASHINGTON, USA

The climate change movement has seen a dramatic shake-up in recent years with the birth of a primarily student- and youth-led movement to remove investments in the fossil fuel industry. While the demands of the fossil fuel divestment movement vary, the most common aim is to encourage institutions to divest from the Carbon Underground 200 – a list which identifies the top 100 public coal companies globally and the top 100 public oil and gas companies globally, ranked by the potential carbon emissions content of their reported reserves. Other campaigns, such as those of the universities of Oxford and Washington, have targeted the most carbon- and capital-intensive fossil fuels such as coal and oil or tar sands. At the core of all such action is a recognition of and calling out of the unsustainable and harmful business model of the fossil fuel industry.

Beginning in spring 2010 on the campus of Swarthmore College, the student group Swarthmore Mountain Justice formed a coal divestment campaign in solidarity with frontline communities fighting against mountaintop removal coal mining in Appalachia (Grady-Benson and Sarathy, 2015). In autumn 2011, Unity College became the first university to divest. By the end of 2013 – just over two years after the movement began – more than 70 institutions had committed to divest. By September 2014, the number of commitments had more than doubled to 181 entities and 650 individuals with control over approximately US$50 billion in total assets (Arabella Advisors, 2014). Representing a fivefold increase, by September 2015 over 442 institutions representing US$2.6 trillion worth of assets, including the Norwegian Sovereign Wealth Fund, the University of Oxford and the Church of England, had all announced plans to divest themselves of some or all of their fossil fuel holdings (Arabella Advisors, 2015). In the run up to COP 21 in Paris more than 100 additional institutions, controlling $US800 billion, committed to divest – bringing the total to $3.4 trillion.

The fossil fuel divestment movement (FFDM) is inspired by a powerful history of students calling for institutional investments to match the values of those institutions. Most prominently, it is modelled after the South African apartheid divestment movement, which asked for institutions to divest from companies operative in apartheid South Africa (see Massie, 1997). In October 2013, Oxford university's Stranded Assets unit released a report examining the history of divestment movements and illustrating the different waves in which they typically occur (see Ansar et al., 2013). In the first wave of divestment, small US institutions divest and so begin to shift the tide of public opinion. In the second wave, bigger, more prestigious US institutions divest. The second wave typically marks a tipping point which triggers the third wave of divestment, global divestment and, potentially, the wide-scale shifting of social and market norms. According to the report, the increasing numbers of people trying to get investments out of the fossil fuel industry represents the fastest growing movement of its kind in history. Reflecting on the progress of the FFDM since the report, the movement is growing so fast that it has arguably already entered the second and third wave of divestment, i.e. entailing the shifting of market and social norms and divesting by major international pension funds and universities – progress it took other divestment movements many more years to achieve.

While keeping track of the size of the movement is difficult, in the space of just four short years the divestment movement has grown from a single campaign on a small college campus in the US to over 400 campaigns in the US and over 800 in total globally in places as diverse as South Africa, Australia, New Zealand, India, Bangladesh, the United Kingdom, European countries, the Marshall Islands, Canada and more (Grady-Benson and Sarathy, 2015). While the campaign is still largely student- and youth-driven, it has garnered high-profile support from the United Nations (UN), the World Bank, the United Nations Framework Convention on Climate Change (UNFCCC) and many more (King, 2015a). The profile, size and rate of divestment is rapidly increasing, and not only is the fossil fuel divestment movement helping to draw attention to the moral urgency of acting on climate change and moving beyond fossil fuels, it is also drawing significant public attention to the carbon bubble and the financially unsustainable nature of the fossil fuel industry's collective and individual business models (King, 2015b).

As a result of the movement's growth both in and beyond institutions of higher education, the FFDM has arguably had significant influence on the public, the financial industry and climate change movements at large. Of course, attributions of causal influence are difficult in such a complex space with multiple different factors at play, but as the next sections hope to illustrate, the FFDM has arguably played a major role in shifting ways of thinking around climate change, enabling coalition building, and has expanded its influence well beyond college campuses and into the international realm.

Table 1. Meadows' places to intervene in a system (in increasing order of effectiveness)

12	Constants, parameters, numbers (such as subsidies, taxes, standards)
11	The size of buffers and other stabilising stocks, relative to their flows
10	The structure of material stocks and flows (such as transport network, population age structures)
9	The lengths of delays, relative to the rate of system change
8	The strength of negative feedback loops, relative to the impacts they are trying to correct
7	The gain around driving positive feedback loops
6	The structure of information flow (who does and does not have access to what kinds of information)
5	The rules of the system (such as incentives, punishment, constraints)
4	The power to add, change, evolve, or self-organise system structure
3	The goals of the system
2	The mind-set or paradigm out of which the system – its goals, structure, rules, delays, parameters – arises
1	The power to transcend paradigms

Source: Authors' elaboration based on Meadows (2009).

Impacts of fossil fuel divestment

Systems thinking may help shed light on the importance of the fossil fuel divestment movement. Applying this thinking to climate change, our system in this case is world society interacting with the planet's climate. Within this system, it is clear that human activity – particularly the burning of fossil fuels – is largely responsible for current observed warming trends in the earth's atmosphere (IPCC, 2014). However, it is also clear that actions in response to this warming – current policies and individual lifestyle changes – are not sufficient to prevent dangerous global warming (Peeters et al., 2015). The FFDM is helping to push past this inaction and the limited analysis and paradigms that underpin it. To better understand how, we can turn to Donella Meadows (2009), who addresses 12 ways to intervene in a system so as to create feedback loops and larger-scale change (Table 1). Among those leverage points are: numbers, constants and parameters (number 12); and the paradigms within which the system functions (numbers 1 and 2); both of which can be found at the heart of the FFDM. Moreover, the rules and structure of the system (numbers 5 and 6) are also being affected, as fossil fuel divestment campaigns aim to empower young people, shift power dynamics and stigmatise and penalise corporations that are obstructing progress on climate change.

The divestment maths

In his 'Do the Math' speaking series, which helped launch the fossil fuel divestment movement, Bill McKibben (2012) addresses three main numbers which illustrate the unsustainable nature of the fossil fuel industry business model and which are at the core of the FFDM:

2°C The maximum amount of total global temperature rise that the Intergovernmental Panel on Climate Change stated can occur before serious feedback loop thresholds – such as ocean acidification and permafrost thaw methane release – will likely begin. This is also the maximum amount of total average global temperature rise that UNFCCC countries have committed to as part of the COP15 agreement in Copenhagen, Denmark, in 2009.

565 $GtCO_2$ (Gigatons of Carbon Dioxide) → The maximum amount of carbon dioxide, according to the Carbon Tracker Institute, that can be emitted in order to stay within the two degrees Celsius warming limit.

2,795 $GtCO_2$ The amount of carbon dioxide already contained in the proven coal and oil and gas reserves of the fossil fuel industry – including both state- and privately owned companies.

The contradiction between the amount we can afford to burn if we are to avoid climate destabilisation and the reserves held by the fossil fuel industry is significant. Indeed, if governments regulate fossil fuels in line with the two degrees target, then two thirds to four fifths of the reserves that fossil fuel companies count as assets on their balance sheet may not be monetised and could become stranded assets (IPCC, 2014). The amount stranded depends on how high of a chance of staying below two degrees is acted on alongside other factors such as the feasibility of carbon capture and storage (see CTI, 2013).

The above contradiction is worrying not only for our climate but also for financial markets. The pioneering work of institutions like the Carbon Tracker Initiative (CTI), HSBC Bank and others have illustrated that this contradiction could potentially result in a systematic overvaluation of fossil fuel companies (CTI, 2012; 2013). Furthermore, not only is the value of publicly traded fossil fuel based on their current reserves, the majority of which may not be burnable, they are also expending approximately 1 per cent of global gross domestic product (GDP) on developing new reserves (CTI, 2013) – ironically this is about the same amount required to invest in the clean economy in order to stay below the 2°C target (see Stern, 2007; IEA, 2014).

Building on their work, it has been estimated that staying below the two degrees target could result in losses of revenue for the fossil fuel industry of up to if not more than US$28 trillion in the next two decades, and potentially over US$100 trillion by 2050 (Lewis, 2014; Channell et al., 2015). The drivers behind fossil fuel companies losing value are not limited solely to climate regulation, but include the changing economics of clean energy, increasing costs of fossil fuel extraction, possible litigation, and a range of other factors (see Paun et al., 2015). The potential risk this poses to investors has been referred to as carbon risk, which forms part of a potentially systematic overvaluation on the financial markets. This has been referred to as the carbon bubble.

As such, while the fossil fuel divestment movement is modelled on other successful divestment campaigns, there are significant differences. Unlike other divestment campaigns, alongside its ethical and political motivations, divestment from the fossil fuel industry has a significant financial motivation associated with potential systematic overvaluation of fossil fuel companies on financial markets and the potential to align our investments with a relatively safe climate future.

Reflective of the power of numbers to shift paradigms, the numbers and figures underpinning the carbon bubble have been at the core of the case for fossil fuel divestment, demonstrating that the overwhelming majority of known fossil fuel reserves must be left in the ground in order to meet the challenges of global warming. This realisation has been at the centre of the Leave it in the Ground (LINGO) movement, which has been motivating a transition to renewable energy and reduced dependence on fossil fuels. Perhaps most significantly, LINGO has helped move the conversation on climate change from a standpoint of reducing emissions to a focus on the need to leave fossil fuels in the ground, unburned. While we cannot say that the current prominence of LINGO is due solely to the FFDM, particularly as it is a concept that predates the campaign, the movement has certainly helped bring it to light (see Bond 2011; Klein, 2014). For instance, the Guardian Media Group's focus on climate change, and their subsequent *#KeepItInTheGround* campaign, has been inspired by campus fossil fuel divestment campaigns (Rusbridger, 2015). Beyond *The Guardian*, the movement has garnered an exponentially growing amount of media coverage, with thousands of divestment-related stories being published in media outlets since the movement began.

On a deeper level, there have been shifts in the paradigm thinking around responsibility for climate change. Previously, responsibility for climate change has been predominantly attributed to either individuals (Peeters et al., 2015) or countries at large, and the organisations – such as the United Nations Framework Convention on Climate Change – that represent them (Kutney, 2014). The FFDM has helped broaden the scope of responsibility to include the role that the fossil fuel industry has played in both contributing to and blocking progress on climate change. It has helped establish an environment where it is possible to talk about the morality of fossil fuel corporations, exposing the unsustainable nature of their business model, highlighting their tactics of deception and misinformation, and in doing so revoking their moral and social license (Mulvey et al., 2015). This has particular significance for institutions of higher education that have not only had their scientific products undermined by fossil fuel industry misinformation – increasingly the economics of clean energy and the feasibility of a transition are being undermined by industry disinformation. In the words of Paul Krugman (2015), 'old Energy is engaged in a systematic effort to blacken the image of renewable energy, one that closely resembles the way it has supported "experts" willing to help create a cloud of doubt about climate science'.

It is through revoking the moral and social license of the fossil fuel industry that divestment activists may have some of their more profound effects. As a University of Oxford Stranded Assets Programme report illustrates (Ansar et al., 2013): through stigmatisation of targeted industries 'almost every divestment campaign [...] from adult services to Darfur, from tobacco to South Africa, divestment campaigns were successful in lobbying for restrictive legislation'. Similarly, the social and political power that the FFDM brings to bear seems set to spur on significant restrictive legislation, if it has not had a role in doing so already. This is especially prominent when coupled with considerations of how the divestment movement is a training ground for much broader climate justice work and serves as a solidarity network and partner in the broader climate justice movement.

Impacts of fossil fuel divestment movement on coalition building

Fossil fuel divestment has often paved the way for coalition building, both within the movement and among other justice groups. On one level, the FFDM allows campus campaigns to directly connect with communities on the front line of fossil fuel extraction and infrastructure development. For example, many campus divestment chapters have organised events and collaborations in solidarity with carbon pricing initiatives and campaigns against mountaintop removal coal mining (MTR), Arctic drilling and hydraulic fracturing (fracking) (Aronoff and Maxmin, 2014). Beyond environmental justice, many divestment groups have organised events in collaboration with the Black Lives Matter movement, and are working on the intersection of racial and climate justice (Divest Harvard, 2014). Other groups, such as Divest Columbia for a Fossil Free Future, have worked with campus prison divestment campaigns. Furthermore, fossil fuel divestment campaigns have created space for network building among campuses themselves as evidenced by the Divestment Student Network.

Other FFDM benefits are oftentimes overlooked. For example, it is also a training ground for new and young climate justice organisers, thus growing the climate change movement at large. The FFDM trains many student organisers *en masse*, oftentimes attracting students who are not already engaged in environmental activism. Thus, a new generation of climate leaders is emerging. Moreover, the FFDM also creates a network within which larger-scale events may be organised. An example of such was the XL DISSENT campaign, a youth-led nonviolent civil disobedience against the Keystone XL pipeline that was held in March 2014. The event, which brought together hundreds of young people to Washington, DC, was organised primarily by students organising for fossil fuel divestment at universities across the country (Democracy Now, 2014). The networks and coalitions created by campus campaigns facilitated the organising of the event.

International impacts of fossil fuel divestment

Power dynamics are at the core of understanding climate change and climate justice. Inherent to justice issues are different realms of power that affect and disempower different social groups (Foucault, cited in Rabinow, 1984). Typically speaking, elites, in an effort to maintain the status quo, 'supply other people with the mental frames for understanding, interpreting and interacting with the world' (Malitz, 2012). This realm of power (Foucault, cited in Rabinow, 1984) is one in which the group being disempowered is often not fully aware of the direct way in which power is exerted upon them. This is particularly important in the case of climate change, for as Kofler and Netzer

(2014) point out, 'only rarely are there immutable facts or technical conflicts that impede or even prevent the expansion of renewable energy. Instead, long-established structures and elites problematise the challenges of an energy transformation and sustain the existing system and their own (market) power with corresponding narratives'.

In the face of elite narratives holding back a clean energy transition, the fossil fuel divestment movement has helped shift those power dynamics and empower young people who have frequently felt disempowered within climate change discussions (Langholz, 2015; Pilrainen, 2015; Yona, 2015). By leveraging the influence of centres of power and credibility such as religious, academic and charitable institutions, the FFDM is providing authority to the voice of young activists and challenging the narrative of the fossil fuel industry on clean energy and the prospects of tackling the climate crisis. When previously youth had been tokenised within spaces like the UNFCCC and refused a seat at the table, the FFDM has helped empower them both within and outside of these spaces and provided them with a louder voice, catapulting the local conversations about fossil fuel divestment that were being had at university campuses to the international stage. For instance, the FFDM is challenging the transnational powers of multinational corporations within the UNFCCC. The COP20 in Lima, Peru (2014) is an example of how the FFDM helped question the legitimacy of fossil fuel corporations' presence at the UNFCCC talks. At the summit, the Global Carbon Capture and Storage Institute organised a side event initially titled, 'Why Divest When a Future with Low Emission Fossil Energy Use is Already a Reality?' The event rapidly drew widespread condemnation from civil society groups present at the conference (Henn, 2014). Its name subsequently changed numerous times, omitting the mention of fossil fuel divestment. These changes still did not prevent media attention, particularly since many divestment campaigns utilised their social media presence to draw attention to the event; on the day of the event, protest actions were organised within the UN convention centre, leading to a significant blockade of the event (DeMelle, 2014).

Many activists drew parallels between the ethics and legitimacy of allowing polluters into a climate change conference with the presence of tobacco companies at World Health Organisation (WHO) summits, arguing that, like tobacco companies, fossil fuel companies could not be part of the solution (Henn, 2014; Supran and Achakulwisut, 2014). Such developments contrasted with the previous year's COP19 in Warsaw, Poland, where a summit of the World Coal Association took place next to the UNFCCC convention venue. At the World Coal Association summit, UNFCCC secretary general Christiana Figueres agreed to be a keynote speaker, despite protests by civil society groups (The Global Call for Climate Action, 2013). Contrasting the two occasions, the events that took place at COP20 represent a different industry response to civil society pressure compared to past COP interactions: whereas at COP19, fossil fuel industry events took place without a hitch, at COP20 the industry came under significant scrutiny.

Due to the rapid turnover of university students, a growing number of alumni have worked on fossil fuel divestment campaigns. Many of these alumni have continued to engage in and organise for climate change – for example Morgan Curtis, a recent graduate of Dartmouth College, is now a youth delegate for the COP21 United Nations climate change negotiations in Paris. Similarly, Kyle Murphy, co-founder of Divest University of Washington, is now a co-director of Carbon Washington, a non-profit running a carbon tax ballot initiative in Washington State. Teddy Smyth, a campaigner for the Middlebury College divestment campaign, now works for Next Gen Climate New Hampshire, an organisation dedicated to making climate change an electoral issue. The list of fossil fuel divestment activists now working on climate justice could certainly go on.

The FFDM also engages older alumni in its effort to sway university trustees and administrations. Examples of these decision-making alumni include Christiana Figueres, secretary-general of the UNFCCC, who recently endorsed fossil fuel divestment from Swarthmore College, her alma mater (Goldenberg, 2015). Others include civil rights activist Cornel West and actor Natalie Portman, who both urged Harvard to divest from fossil fuels (Klein, 2015). Beyond these influential alumni, many college fossil fuel divestment campaigns have created escrow funds that allow alumni to divert their donations to a fund that will only be made accessible once the alma mater divests. Additionally, campaigns at a couple of dozen institutions – including Boston University and Dartmouth College – have collaborated in a Multi-School Divest Fund, which allows alumni to pool their resources together in an escrow fund. By December 2017, the funds received will be divided equally, but only among the schools that have divested (Rocheleau, 2015). Through its engagement with alumni, the FFDM is helping to spread the divestment and LINGO movement well beyond campuses and into communities across the globe, often in powerful and influential places.

Future trajectories

Looking forward, the divestment movement arguably has growing potential to help bolster the broader climate justice movement and spur on the enactment of climate policy. This is of particular significance given that FFDMs took off primarily within developed countries who have typically been perceived as laggards within the UNFCCC space, such as the United States, Australia and Canada; countries which arguably have a strong moral responsibility to act under the principle of Common but Differentiated Responsibilities, central to the UNFCCC and most accounts of climate justice. In these countries, the FFDM is helping to provide the political and social space required to enact policies like carbon pricing, fossil fuel subsidy reform, and more. Moreover, we expect the FFDM to play a growing role in helping promote the conversation around ·decarbonisation and the push for 100 per cent renewable energy by 2050 as it continues to grow (Sachs, 2015).

Looking forward, the divestment movement arguably has growing potential to help bolster the broader climate justice movement and spur on the enactment of climate policy. This is of particular significance given that FFDMs took off primarily within developed countries who have typically been perceived as laggards within the UNFCCC space, such as the United States, Australia and Canada; countries which arguably have a strong moral responsibility to act under the principle of Common but Differentiated Responsibilities, central to the UNFCCC and most accounts of climate justice. In these countries, the FFDM is helping to provide the political and social space required to enact policies like carbon pricing, fossil fuel subsidy reform, and more. Moreover, we expect the FFDM to play a growing role in helping promote the conversation around decarbonisation and the push for 100 per cent renewable energy by 2050 as it continues to grow (Sachs, 2015).

Furthermore, while many activists are (perhaps too) quick to grant that divestment is not about directly bankrupting the fossil fuel industry, the influence on the investment world is increasingly evident and rapidly growing. As both the University of Oxford Stranded Assets Programme (Ansar et al., 2013) and Peabody Coal have realized, divestment efforts 'could significantly affect demand for [fossil fuel] products' and could 'indirectly influence all investors [...] to go underweight on fossil fuel stocks and debt in their portfolios'. It seems, furthermore, that such impacts are already with us. For instance, as Shawn McCarthy (2015) reports, the FFDM has already 'grown into something much larger and more threatening to producers [than simply a political movement], as pension fund managers and other institutional investors are now questioning the long-term returns offered by coal and oil companies'.

Following in the footsteps of a number of financial institutions, HSBC Bank recently issued a research report warning investors that the fossil fuel industry is at serious and growing risk of stranded assets from climate policies and unfavourable economics. In their report, HSBC suggested a number of different divestment strategies and argued that divestment could affect fossil fuel production and extend the carbon budget by creating 'less demand for shares and bonds [which] ultimately increases the cost of capital to companies and limits the ability to finance expensive projects, which is particularly damaging in a sector where projects are inherently long term' (Paun et al., 2015).

If HSBC and a number of analysts are correct, the FFDM will continue to grow and help to shift significant amounts of capital. Such a shift is urgently needed, for as the financial analysts at the 2° Investing Initiative (2013) have pointed out, 'divesting from fossil fuels is an integral piece to aligning the financial sector with a 2°C climate scenario'. Their claim is substantiated by the International Energy Agency (IEA), which estimates that alongside a major increase in clean energy investments, reductions in fossil fuel investments of US$4.9 trillion and additional divestment away from fossil-fuelled power transmission and distribution of US$1.2 trillion will be needed by 2035 if we are to achieve the internationally agreed-upon 2°C target. Furthermore, because we are seeing the FFDM expand to target banks, foundations, pension funds, and financial institutions with growing amounts of success, the movement is increasingly helping to shift much greater amounts of capital.

While the divestment movement began with just a few small campaigns on campuses in the US, the movement has grown significantly and its impacts and reach are increasingly vast and global in nature. Of course, divestment efforts when viewed in isolation are seemingly insufficient to the task of addressing the climate crisis as they do not directly address many facets of what is required to tackle the crisis. However, as we have highlighted, when viewed in the broader context of the growing climate justice movement, the FFDM has been a powerful force multiplier, ally and partner, which has helped propel issues such as LINGO, climate justice and the carbon bubble into the foreground of public consciousness and which can help spur on some of the requisite legislation needed to leave fossil fuels in the ground. Thus, while the FFDM is by no means the sole solution to the climate crisis, it is increasingly proving that it is an important component and driver of the broader alliance of civil society, politics, science, and industry who are developing a convincing alternative and positive narrative to counter the fossil fuel industry narrative and implementing them against resistance.

References

2° Investing Initiative (2013). *From Financed Emissions to Long Term Investing Metrics: State of the Art Review of GHG Emissions Accounting for the Financial Sector.* Retrieved from http://www.trucost.com/published-research/112/2i

Ansar, A., Caldecott, B. and Tilbury, J. (2013). *Stranded Assets and the Fossil Fuel Divestment Campaign : What Does Divestment Mean for the Valuation of Fossil Fuel Assets ?* Oxford: SSEE-University of Oxford. Retrieved from http://www.bsg.ox.ac.uk/stranded-assets-and-fossil-fuel-divestment-campaign-what-does-divestment-mean-valuation-fossil-fuel

Arabella Advisors (2014). *Measuring the Global Fossil Fuel Divestment Movement.* Retrieved from http://www.arabellaadvisors.com/wp-content/uploads/2014/09/Measuring-the-Global-Divestment-Movement.pdf

Arabella Advisors (2015). *Measuring the Growth of the Global Fossil Fuel Divestment and Clean Energy Investment Movement.* Retrieved from http://www.arabellaadvisors.com/wp-content/uploads/2015/09/Measuring-the-Growth-of-the-Divestment-Movement.pdf

Aronoff, K. and Maxmin, C. (2014, 9 April). A Generation's Call: Voices from the Student Fossil Fuel Divestment Movement. *Dissent Magazine.* Retrieved from https://www.dissentmagazine.org/online_articles/a-generations-call-voices-from-the-student-fossil-fuel-divestment-movement

Bond, P. (2011). *Politics of Climate Justice: Paralysis Above, Movement Below.* Pietermaritzburg: University of KwaZulu-Natal Press.

Channell, J., Jansen, H. R., Curmi, E., Rahbari, E., Nguyen, P., Morse, E. L., Prior, E., Kleinman, S. M., Syme, A. R. and Kruger, T. (2015). *Energy Darwinism II: Why a Low Carbon Future Doesn't Have to Cost the Earth.* Citi GPS: Global Perspectives & Solutions. Retrieved from https://ir.citi.com/E8%2B83ZXr1vd%2Fqyim0DizLrUxw2FvuAQ2jOlmkGzr4ffw4YJCK8s0q2W58AkV%2FypGoKD74zHfji8%3D

CTI (2012). *Unburnable Carbon – Are the World's Financial Markets Carrying A Carbon Bubble?* London. Carbon Tracker Initiative. Retrieved from http://www.carbontracker.org/wp-content/uploads/2014/09/Unburnable-Carbon-Full-rev2-1.pdf

CTI (2013). U*nburnable Carbon 2013: Wasted Capital and Stranded Assets*. London. Retrieved from http://carbontracker.live.kiln.it/Unburnable-Carbon-2-Web-Version.pdf

DeMelle, B. (2014, 10 December). Lima COP20 Climate Summit Marked by Protests, Publicity Stunts, and PR Promises. *DeSmog Blog*. Retrieved from www.desmogblog.com/2014/12/11/lima-cop20-climate-summit-marked-protests-promises-and-publicity-stunts

Democracy Now (2014, 3 March). *XL Dissent: 398 Youth Arrested at Anti-Keystone XL Pipeline Protest at White House*. Retrieved from www.democracynow.org/2014/3/3/xl_dissent_398_youth_arrested_at

Divest Harvard. (2014, 15 December). *Divest Harvard Statement on Black Lives Matter*. Retrieved from divestharvard.com/divest-harvard-statement-on-black-lives-matter/

Goldenberg, S. (2015, 24 March). UN Climate Chief Joins Alumni Calling on Swarthmore to Divest from Fossil Fuels. *The Guardian*. Retrieved from http://www.theguardian.com/environment/2015/mar/24/un-climate-chief-joins-alumni-calling-swarthmore-college-divest-fossil-fuels

Grady-Benson, J. and Sarathy, B. (2015). Fossil fuel divestment in US higher education: student-led organising for climate justice. *Local Environment: The International Journal of Justice and Sustainability*. Retrieved from http://www.tandfonline.com/doi/full/10.1080/13549839.2015.1009825

Henn, J. (2014, 5 December). A Fossil Fuel Scandal at the Climate Talks in Lima. *Huffington Post*. Retrieved from www.huffingtonpost.com/jamie-henn/a-fossil-fuel-scandal-at-_b_6278018.html

IEA (2014). *Energy Technology Perspectives 2014: Harnessing Electricity's Potential*. Paris. Retrieved from http://www.iea.org/bookshop/472-Energy_Technology_Perspectives_2014

IPCC (2014). *Summary for Policymakers*. In: IPCC. Climate Change 2014: Mitigation of Climate Change. Contribution of Working Group III to the Fifth Assessment Report of the Intergovernmental Panel on Climate Change. Retrieved from http://mitigation2014.org/

King, E. (2015a, 23 January). World Bank Chief Backs Fossil Fuel Divestment Drive. Climate Home. Retrieved from http://www.rtcc.org/2014/01/27/world-bank-chief-backs-fossil-fuel-divestment-drive/

King, E. (2015b, 18 August). Terrifying Math: How Carbon Tracker Changed the Climate Debate. *Climate Home*. Retrieved from http://www.rtcc.org/2015/08/18/terrifying-math-how-carbon-tracker-changed-the-climate-debate/

Klein, M. A. (2015, 20 February). Divest Harvard Plans Weeklong Sit-In. *The Harvard Crimson*. Retrieved from www.thecrimson.com/article/2015/2/20/divestment-plans-protest-week/

Klein, N. (2014). T*his Changes Everything: Capitalism Vs. The Climate*. New York: Simon & Schuster Paperbacks.

Kofler, B. and Netzer, N. (eds) (2014). *Towards a Global Energy Transformation*. Friedrich Ebert Stiftung. Retrieved from http://library.fes.de/pdf-files/iez/10817.pdf

Kutney, G. (2014). *Carbon Politics and the Failure of the Kyoto Protocol*. London: Routledge.

Krugman, P. (2015, 5 October). Enemies of the Sun. *The New York Times*. Retrieved from http://www.nytimes.com/2015/10/05/opinion/paul-krugman-enemies-of-the-sun.html?_r=0

Langholz, M. (2015, 27 June). How We Move From Access to Influence. *SustainUS: U.S. Youth for Justice and Sustainability*. Retrieved from sustainus.org/2015/06/how-we-move-from-access-to-influence/

Lewis, M. C. (2014). *Stranded Assets, Fossilised Revenues*. Kepler Cheuvreux. ESG Research. Retrieved from http://www.keplercheuvreux.com/pdf/research/EG_EG_253208.pdf

Malitz, Z. (2012). Intellectuals and Power. In: A. Boyd (ed.). *Beautiful Trouble: A Toolbox for Revolution* (pp. 240-241). New York: OR Books. Retrieved from http://monoskop.org/images/6/67/Boyd_Andrew_ed_Beautiful_Trouble_A_Toolbox_for_Revolution.pdf

Massie, R. K. (1997). *Loosing the Bonds: The United States and South Africa in the Apartheid Years*. New York: Doubleday.

McCarthy, S. (2015, 24 September). Divestment Efforts Starting to Hit Coal and Oil Firms. *The Globe and Mail*. Retrieved from http://www.theglobeandmail.com/report-on-business/industry-news/energy-and-resources/divestment-efforts-starting-to-hit-coal-and-oil-firms/article26535044/

McKibben, B. (2012, 19 July). Global Warming's Terrifying New Math. *Rolling Stone*. Retrieved from http://www.rollingstone.com/politics/news/global-warmings-terrifying-new-math-20120719

Meadows, D. (2009). Leverage Points: Places to Intervene in a System. *The Solutions Journal*, 1(1), 41-49. Retrieved from http://www.thesolutionsjournal.com/node/419

Mulvey, K., Shulman, S., Anderson, D., Cole, N., Piepenburg, J. and Sideris, J. (2015). *The Climate Deception Dossiers: Internal Fossil Fuel Industry Memos Reveal Decades of Corporate Disinformation*. Cambridge, MA: Union of Concerned Scientists. Retrieved from http://www.ucsusa.org/sites/default/files/attach/2015/07/The-Climate-Deception-Dossiers.pdf

Paun, A., Knight, Z. and Chan, W.-S. (2015). Stranded Assets: *What Next? How Investors Can Manage Increasing Fossil Fuel Risks*. HSBC Global Research. Retrieved from http://www.businessgreen.com/digital_assets/8779/hsbc_Stranded_assets_what_next.pdf

Peeters, W., De Smet, A., Diependaele, L. and Sterckx, S. (2015). *Climate Change and Individual Responsibility: Agency, Moral Disengagement, and the Motivational Gap*. Basingstoke: Palgrave MacMillan.

Pilrainen, P. (2015, 14 June). The Abyss and Salvation. *UK Youth Climate Coalition*. Retrieved from ukycc.org/the-abyss-and-salvation/

Rabinow, P. (ed.) (1984). *The Foucault Reader*. New York: Pantheon Books.

Rocheleau, M. (2015, 20 February). Prominent Alumni Ramp Up Pressure on Universities to Divest. *Boston Globe*. Retrieved from https://www.bostonglobe.com/metro/2015/02/19/alumni-withhold-donations-join-student-protests-pressure-colleges-divest-from-fossil-fuels/WgWZ1SQKEAxigN6Gl1pRrl/story.html

Rusbridger, A. (2015, 16 March). The Argument for Divesting from Fossil Fuels is Becoming Overwhelming. *The Guardian*. Retrieved from www.theguardian.com/environment/2015/mar/16/argument-divesting-fossil-fuels-overwhelming-climate-change?CMP=share_btn_tw

Sachs, J. D. (2015, 9 October). Our Era's Moonshot: Deep Decarbonization. *Shanghai Daily*. Retrieved from http://www.shanghaidaily.com/opinion/foreign-perspectives/Our-eras-moonshot-deep-decarbonization/shdaily.shtml

Stern, N. (2007). *The Economics of Climate Change: The Stern Review*. Cambridge: Cambridge University Press.

Supran, G. and Achakulwisut, P. (2014, 12 December). Fossil Fuel Companies Grow Nervous as Divestment Movement Grows Stronger. *Grist*. Retrieved from www.grist.org/climate-energy/fossil-fuel-companies-grow-nervous-as-divestment-movement-grows-stronger/

The Global Call for Climate Action (2013, 8 November). *Global Youth Call on UNFCCC Head to Cancel Coal Summit Attendance*. Retrieved from tcktcktck.org/2013/11/global-youth-call-unfccc-head-cancel-coal-summit-attendance/

Yona, L. (2015, 1 September). Arctic Climate Summit: Do Leaders Love Their Children Enough? *Climate Home*. Retrieved from www.rtcc.org/2015/09/01/arctic-climate-summit-do-leaders-love-their-children-enough/

14

Investing in the Future: Norway, Climate Change and Fossil Fuel Divestment

MATTHEW RIMMER

QUEENSLAND UNIVERSITY OF TECHNOLOGY, AUSTRALIA

The fossil fuel divestment movement has undergone explosive growth over the last few years – expanding from encouraging educational institutions to adopt ethical investment policies to focusing upon cities, pension funds and philanthropic charities. The fossil fuel divestment movement has attained global ambitions – challenging sovereign wealth funds and national governments to engage in fossil fuel divestment, and pushing for fossil fuel divestment at international climate talks – such as the Paris Climate Summit in 2015.

By exploring and analysing a key campaign to 'Divest Norway', this chapter considers the efforts to globalise and internationalise the fossil fuel divestment campaign. Part 1 explores the origins of the fossil fuel divestment movement, and the application of such strategies in a variety of contexts. Part 2 looks at the campaign to divest Norway's sovereign wealth fund of fossil fuel investments. There has been much discussion as to whether the bold decision of Norway to engage in coal divestment will encourage and inspire other sovereign wealth funds to engage in fossil fuel divestment. The conclusion considers the efforts to introduce fossil fuel divestment as a policy initiative for nation states as a policy option in international climate law.

Fossil fuel divestment

The movement had its origins in Vermont academic and philosopher Bill McKibben (2013) and the climate network, 350.Org, calling for fossil fuel divestment by schools and universities. The anti-apartheid campaigner, Bob Massie (2012), provided significant inspiration for the fossil fuel divestment movement. Massie's anti-apartheid divestment strategy had proved to be an effective means of galvanising student and staff support against South Africa's Apartheid Policies. Massie advised Bill McKibben: 'Given the severity of the climate crisis, a comparable demand that our institutions dump stock from companies that are destroying the planet would not only be appropriate but effective [...] We must sever the ties with those who profit from climate change – now' (McKibben, 2013: 152). Appalachian activists against mountaintop removal also provided inspiration for the fossil fuel divestment movement. As Naomi Klein noted, the divestment movement 'emerged organically out of various Blockadia-style attempts to block carbon extraction at its source – specifically, out of the movement against mountaintop removal coal mining in Appalachia, which were looking for a tactic to put pressure on coal companies that had made it clear that they were indifferent to local opinion'. (Klein, 2014: 353)

'Do the Math'

As part of a 'Do the Math' tour, Bill McKibben, 350.Org and 'Go Fossil Free' promoted a movement to encourage divestment in fossil fuel industries. McKibben recognised that movements rarely have predictable outcomes. However, he maintained that a 'campaign that weakens the fossil-fuel industry's political standing clearly increases the chances of retiring its special breaks' (2012).

Naomi Klein (2014; 2015) considered that 'another tactic spreading with startling speed is the call for public interest institutions like colleges, faith organizations, and municipal governments to sell whatever financial holdings they have in fossil companies'. The fossil fuel divestment movement initially focused upon schools and universities. Klein (2014) noted: 'Young people have a special moral authority in making this argument to their school administrators: these are the institutions entrusted to prepare them for the future; so it is the height of hypocrisy for those same institutions to profit from an industry that has declared war on the future at the most elemental level.'

She also observed that the strategy is designed to remove social respectability from fossil fuel companies: 'The eventual goal is to confer on oil companies the same status as tobacco companies, which would make it easier to make other important demands – like bans on political donations from fossil fuel companies and on fossil fuel advertising on television (for the same public health reasons that we ban broadcast cigarette ads)'. She hopes that there will be 'space for a serious discussion about whether these profits are so illegitimate that they deserved to be appropriated and reinvested in solutions to the climate crisis' (Klein, 2014).

In a further essay, Klein (2015) elaborated that fossil fuel divestment policies had been adopted by Stanford University, Glasgow University, and the Rockefellers. She wondered whether fossil fuel companies – long toxic to the natural environment – became also toxic in the field of public relations. Under pressure from Greenpeace, even Lego ended its long-standing relationship with Shell (Vaughan, 2014). Klein (2015) noted: 'At their core, all are taking aim at the moral legitimacy of fossil fuel companies and the profits that flow from them [...] This movement is saying that it is unethical to be associated with an industry whose business model is based on knowingly destabilising the planet's life support systems.'

City governments and pension funds

In addition to universities, there has also been a focus upon city governments and pension funds (Saxifrage, 2013). Bill McKibben urged pension funds to desist from investment in fossil fuels, observing that 'it does not make sense to invest my retirement money in a company whose business plan means that there won't be an earth to retire on' (Gunther, 2012). In San Francisco, councillor John Avalos proposed that the city's retirement fund should withdraw its money from fossil fuels (Green, 2013). McKibben commented on the proposal: 'The Bay Area will spend billions adapting to climate change – it makes no sense at all to simultaneously invest in the corporations making that work necessary' (350.org, 2013). In the United States (US), a number of progressive cities – such as Seattle (McGinn, 2012) and Portland (Law, 2015) – have supported fossil fuel divestment initiatives. The state legislature of California has passed a coal divestment bill (Carroll, 2015). In Australia, cities such as Fremantle (City of Fremantle, 2014), Canberra (Edis, 2015), Newcastle (Saunders, 2015) and Melbourne (350.org, 2015), towns such as Lismore, and the local councils of Leichhardt, Marrickville and Moreland have all adopted fossil fuel divestment policies. In New Zealand, Dunedin and Christchurch have adopted fossil fuel divestment policies (Free Speech Radio News, 2015). The City of Victoria in Canada has supported fossil fuel divestment (Cleverley, 2015). In Norway, the City of Oslo has also decided to divest from fossil fuels (Agence France-Presse, 2015).

Charities, philanthropies, and religious institutions

A number of charities and philanthropists – including the Rockefeller Foundation, the Children's Investment Fund Foundation, the KR Foundation, Leonardo DiCaprio and the Leonardo DiCaprio Foundation – have embraced fossil fuel divestment (Goldenberg, 2015; Rowling, 2015). There has been a concerted campaign by *The Guardian* newspaper called 'Keep It in the Ground', which has sought to encourage the charities Gates Foundation and the Wellcome Trust to divest from fossil fuels (Rimmer, 2015a). Health professionals around the world have also made the decision to engage in fossil fuel divestment (Picard, 2015). Religious institutions have also been attracted to ethical investment policies (Green, 2014), with organisations such as the World Council of Churches committing to engage in fossil fuel divestment.

The impact of the fossil fuel divestment movement

There has been an increasing amount of scholarship upon the symbolic and the practical impact of the fossil fuel divestment movement. Professor Ben Caldecott and his group at the University of Oxford have undertaken extensive work upon stranded assets. The group contends that 'divestment campaigns will probably be at their most effective in triggering a process of stigmatisation of fossil fuel companies' (Ansar et al., 2014: 74).

In his book, *The Energy of Nations*, Jeremy Leggett (2014: 189) observed that there was growing institutional pressure upon fossil fuel companies to address climate risks: 'Failure to address the threat of unburnable carbon could leave players exposed to material asset write-downs and wasted investment, both potentially destroying shareholder value.' He noted that 350. org and civil society provided additional pressure, demanding that a range of institutions engage in fossil fuel divestment. Leggett hoped: 'With citizen pressure using 350.org's language of morality, and institutional pressure using Carbon Tracker's language of capital, we stand to create a pincer movement' (p. 189).

In his book, *Atmosphere of Hope*, Tim Flannery (2015) – a councillor at the Climate Council – considered the impact of the fossil fuel divestment movement. He observed that 'the recognition that fossil fuel companies are fundamentally overvalued, because most of their assets cannot be used if we are to have a stable climate, has led to investors selling off their shares in various fossil fuel-based industries' (p. 106).

In *The Carbon Bubble*, Jeff Rubin considers the impact of the fossil fuel divestment movement. He noted: 'Given the historically poor performance of ethically motivated investments, the mass exodus is – at least at first glance – surprising' (2015: 161) Rubin observes that there are compelling justifications for fossil fuel divestment: 'These days, then, divestment from carbon has become a win-win scenario, and a somewhat easy move for institutional investors to make.' He commented that, in ethical terms, 'divesting form oil, coal or natural gas looks good in an increasingly enviro-conscientious world' (p.161). Moreover, Rubin observed that, in financial terms, fossil fuel divestment has become attractive, given the risks associated with stranded carbon assets: 'As countries around the world increasingly take measures to clamp down on their carbon emissions, your investment portfolio will yield higher returns if it doesn't hold any coal, oil or gas stocks.' (p.162)

The fossil fuel divestment movement has been extraordinarily influential over a short period of time, with US$ 2.6 trillion dollars in commitments as of September 2015 (Martin, 2015). A report by Arabella Advisors (2015) found that 'divesting from fossil fuels and investing in clean energy has empowered thousands of institutions and individuals across the world to take direct action on climate, as 436 institutions and 2,040 individuals across 43 countries representing US$ 2.6 trillion in assets have pledged to divest' (p. 16). The report observed: 'The increasing likelihood of near term carbon regulation has created financial risks to portfolios exposed to fossil fuel assets, which has driven exponential growth in divestment in new sectors including pension funds and private institutional investors' (Arabella Advisors, 2015: 16) The report found: 'At the same time, mission-driven organizations are making a strong moral case for divestment, as faith communities, universities, health care organizations, and foundations continue to drive remarkable growth in commitments' (p.16). The report stressed: 'Together, they are sending a clear signal that they consider fossil fuel investments too risky in a carbon constrained world' (p.16).

Norway's sovereign wealth fund: ethical investment, renewable energy, and climate change

In 2014 and 2015 there was a significant public debate over whether Norway's sovereign wealth fund should invest in renewable energy; divest from fossil fuels; and engage in ethical investment. At a massive US$840 billion, Norway's sovereign wealth fund owns 1 per cent of all the publicly listed companies in the world with investments spread across more than 8,000 companies in 82 countries. The fund, made up of Norway's surplus tax revenues from oil and gas production, was established in 1990, partly in an effort to manage the impacts of volatile oil prices. In 2014, Norway faced a new challenge – the petroleum resources that its wealth was derived from started to peak. Consequently, Norway is currently in the midst of a comprehensive debate over what to do about its vast petro-wealth, including a review of whether to divest the fund of all coal, oil and gas companies.

In May 2015, the Norwegian government presented plans for a new climate criterion for the exclusion of companies from the Government Pension Fund Global. The Minister of Finance, Siv Jensen, commented: 'The Committee expects the Government to propose a concrete, new product-based criterion in the National Budget for 2016 this autumn and the new criterion to be put in place by 1 January 2016' (Government of Norway, 2015b). She also commented: 'The Government will follow-up the Storting's deliberations, and will as part of its work ask Norges Bank and the Council on Ethics for advice' (ibid.)

The Storting's Standing Committee on Finance and Economic Affairs believed that it was appropriate to introduce a new product-based criterion aimed at mining companies and power products, which had a significant portion of its business and income related to coal. The Committee was of the view that the rule would affect companies that base 30 per cent or more of their activities on coal and/or derive 30 per cent or more of their revenue from coal.

Renewable energy investment

In March 2014, Norway's Prime Minister Erna Solberg announced her government's plans to invest a significant proportion of the nation' sovereign wealth fund in renewable energy in an effort to cut greenhouse gas emissions and address climate change. She noted: 'This government takes environmental problems very seriously but we need to have a good look at how to address through positive investments in renewable energy in sustainable companies overseas through the fund' (Phillips, 2014). Solberg stressed: 'It's important that Norway leads the way beyond our borders (ibid.).'

In an April 2014 speech, Norwegian Finance Minister Siv Jensen announced that the fund would double its investment in renewables to around US$8 billion. She said: 'The increased scope we give on green investments will help the fund's ability to actively manage investments in this area.' However, she warned that the fund 'is not a tool to boost government investments in emerging markets or renewables' (Government of Norway, 2014a). Her sister, WWF Norway CEO Nina Jensen called the changes 'peanuts', stating: 'Norway can make a huge difference in the world...this announcement falls short of meeting expectations of the people of Norway and of the world' (WWF Norge, 2014). She responded, 'every decision Norway makes on this fund sends signals around the world' (ibid).

However, Terje Osmundsen of the Norwegian Climate Foundation called for greater ambition: 'if the fund is allowed to invest up to 5 per cent – equal to the target set for its property investments – of its total assets into renewable energy-related infrastructure, the fund could on average allocate in the range of US$10 billion per year to the green energy investment market from 2015 onwards' (Phillips, 2014). Osmundsen hoped that such an investment could make Norway's sovereign wealth fund one of the world's largest single clean energy investor.

In 2014, clean energy advocates called upon Norway to follow in Denmark's footsteps by committing to invest 5 per cent of the fund in renewable energy infrastructure – a level of investment that they say would be a game-changer for the renewable sector globally. But questions of 1 per cent versus 5 per cent aside, campaigners point out that Norway's green investments mean little while the fund continues to invest in vast quantities of fossil fuels. The case for greater investment in clean technology was strengthened by the United Nations' report on *Mitigation of Climate Change* – which recommended that a huge increase in renewable energy is necessary to avert climate disaster (McKie and Helm, 2014).

The Government of Norway (2015a) noted that the 'report to Parliament announces that the scale of the environment-related investment mandates [...] will be expanded to NOK 30–60 billion', and that 'a process has also been launched to examine whether the Fund should be permitted to be invested in unlisted infrastructure, including renewable energy infrastructure'.

Ethical investment

Norway's sovereign wealth fund excluded a number of companies from the Government Pension Fund, including companies involved in the production of weapons – such as land mines, cluster munitions and the production of nuclear weapons – that violate fundamental humanitarian principles (Government of Norway, 2014b; 2014c). The fund has also banned investment in a company involved in the sale of weapons and military material to Burma, in companies that have contributed – by actions or omissions – to severe environmental damage, in companies involved in the production of tobacco, as well as in those involved in serious or systemic violations of human rights, fundamental ethical norms or individuals' rights in situations of war or conflict.

A number of mining companies have been affected by the bans – including

Sesa Sterlite; WTK Holdings Berhad; Ta Ann Holdings Berhad; Zijin Mining Group; Volcán Compañía Minera; Lingui Development Berhad Ltd; Samling Global Ltd.; Norilsk Nickel; Barrick Gold Corp; Rio Tinto plc; Rio Tinto Ltd; Madras Aluminium Company; Sterlite Industries Ltd; Vedanta Resources plc; and Freeport McMoRan Copper & Gold Inc. Notably, in October 2015, Astra International Tbk PT was placed under observation because of the risk of severe environmental damage (Norges Bank Investment Management, 2015).

The managers of Norway's sovereign wealth fund have also emphasised the need to standardise and enhance global reporting on climate risk. They stressed: 'We expect companies to develop strategies for managing risks related to climate change and report on what they are doing to reduce the risk of climate change impacting negatively on their profitability' (Norges Bank Investment Management, 2013: 43). The ethics of the fund's investments have been a central focus since 2004, when an independent ethics council was established to inform the fund's investment decisions. Under the council's direction, the fund's manager – the Norwegian central bank – has already screened out a number of companies on environmental, health and human rights grounds. As of 1 January 2015, Norges Bank's executive board makes decisions in respect of exclusions – rather than the Ministry of Finance.

Divesting fossil fuels

There has been debate in the Norwegian parliament as to whether Norway's sovereign wealth fund should go further, and divest from coal, oil, and gas. In 2014, Siv Jensen noted: 'Ethical exclusion is a relatively limited tool – as a financial investor we cannot entirely "sell our way" out of potential problems in the investment portfolio' (Government of Norway, 2014a). She also observed: 'Exclusion may also not be the best way to promote change in companies, or to safeguard the financial value of the Fund's investments'. In her view, 'exclusion as it has been used by the Fund is a "measure of last resort" and reserved for the most severe cases.'

In 2014, the Government of Norway established a review, setting up an expert group to evaluate whether excluding investments in coal and oil companies was a more effective strategy for addressing climate issues and promoting future change than the exercise of ownership and exertion of influence. Australian resource companies such as BHP Billiton, Woodside Petroleum and Whitehaven Coal were nervous about the review, because of the significant investments held by Norway's sovereign wealth fund (Ker, 2014). The fund was also a shareholder in Glencore Xstrata, Anglo American,

Shell, ExxonMobil, BP and Chevron.

In May 2015, 'Go Fossil Free' mounted a global campaign to encourage Norway to divest from fossil fuels. The group said: 'It's time for the largest national fund in the world to stop profiting from climate destruction' (Gofossilfree.org, 2015a). That year, the Government of Norway announced that there would be a new climate criterion for the exclusion of companies from the Government Pension Fund Global:

> An expert group has examined the policy tools available to the Fund in relation to climate issues. The report from the group has been circulated for consultation. The Government will introduce, against the background of the report and the consultative comments, a new conduct-based exclusion criterion aimed at *'acts and omissions that, on an aggregate company level, to an unacceptable degree entail greenhouse gas emissions'*. The criterion is broad in scope, and not limited to specific sectors or types of greenhouse gases. It will also accommodate norm changes within this field over time. The wording is identical to that proposed by the Council on Ethics in its consultative comments. Political bodies have adopted ethically motivated criteria for the exclusion of companies from the GPFG [Government Pension Fund Global]. Some of these criteria are based on which *products* companies produce, while others are based on the *conduct* of companies. The intention behind the ethical criteria is to reduce the risk that the Fund is invested in companies that contribute to, or are themselves responsible for, gross violations of ethical norms. (Government of Norway, 2015a)

The Government of Norway agreed with 'the professional assessment of the expert group that ethically motivated exclusion of *all* coal and petroleum companies based on their products would not be appropriate'; and that 'the energy production, energy use or CO_2 emissions of such companies cannot *per se* be said to be contrary to generally accepted ethical norms' (Government of Norway, 2015a). The Government of Norway also noted the opinion of the expert group that using the fund as a climate policy tool would be inappropriate and ineffective. The Minister of Finance, Siv Jensen, said: 'The measures introduced by the Government are premised on the broad consensus concerning the role of the Fund as a financial investor, which has facilitated the robust long-term management of our savings' (Government of Norway, 2015a).

The decision was backed by all the main parties in Norway. Rasmus Hansson, a member of parliament for Norway's Green Party took a stronger view about the significance of the decision: 'We've crossed an important line declaring the fund as a climate policy vehicle' (Mohsin and Holter, 2015). This discussion of the new climate criterion makes it clear that the decision is a partial one in respect of fossil fuel divestment. Nonetheless, the decision has attracted a great deal of public attention, because of the size and scale of the sovereign wealth fund (ABC Environment, 2015; Queally, 2015; Reuters, 2015).

The significance of Norway's decision

There has been a discussion as to whether the decision of the Norwegian sovereign wealth fund will trigger a wave of large fossil fuel divestments by a range of other actors (Carrington, 2015). For example, Mark Campanale, founder of the Carbon Tracker Initiative, said: 'The significance of the Norway decision is that, because of their size and reach, this will act as a major signal for other investors to follow.' Tom Sanzillo, a former comptroller of New York State who oversaw a US$156 billion pension fund, also said Norway's move was likely to spark others to do the same: 'Coal markets globally are in the midst of a wrenching structural decline.' Heffa Schücking, at German NGO Urgewald and who has written several financial reports on Norway's wealth fund, commented: 'This will send a strong signal to investors all over the world. Coal is yesterday's fuel' (Carrington, 2015b).

For all their exuberance, civil society groups were circumspect about the decision. Briefing papers suggested that much will depend upon how the new rules are implemented in Norway (Schücking, 2015). Greenpeace, WWF, Future in our Hands, 350.org and Urgewald maintained: 'Divestment from coal must be the first step for Norway, not the last' (Queally, 2015). The civil society groups said that they would campaign for the fund to invest at least 5 per cent of its value in renewables, particularly in emerging economies, and for full divestment from all fossil fuels; while climate activists maintained that for Norway itself, their goal should be a transition out of oil and gas and into the green jobs of the future. In their view, 'we are rapidly approaching the time when no country can rely on fossil fuels for its economy or energy safety' (Queally, 2015).

In addition to coal divestment, the Norwegian Prime Minister, Erna Solberg, has demanded a global carbon price and an end to fossil fuel subsidies (Shankleman, 2015). In this sense, her neighbour, the Prime Minister of Sweden, Stefan Lovan, has announced that his country will work towards becoming 'one of the first fossil fuel-free welfare states in the world' (Bolton, 2015).

Hopefully, Norway's sovereign wealth fund will lead the way for other major sovereign wealth funds, pension funds, development banks and governments to cut their investments in fossil fuels. Benjamin Richardson (2013a; 2013b) has suggested that Norway's sovereign wealth fund could 'help institutionalize the principles of intergenerational equity and sustainable development in the context of financial markets'.

Considering the rapid developments in respect of fossil fuel divestment, Charlotte Wood of 350.org Australia observed that such decisions were only the beginning and that the end of coal is a 'a reality that's gaining growing acceptance from High Street to Wall Street as investors divest billions of dollars from this dirtiest of fossil fuels' (Wood, 2015). She noted, though, that 'the end of coal won't solve the climate crisis'; commenting that 'if we want a liveable planet, we need to get off all fossil fuels, oil and gas included'. Moreover, Wood highlighted the financial case for fossil fuel divestment: 'Aside from its devastating impacts on our health, climate and communities, most coal stocks have tanked so low that they're bad investments on financial grounds alone'. She noted: 'Smart investors like Stanford University, with their $18bn endowment, and the Norwegian Sovereign Wealth Fund, with its $900bn, are getting their money out of coal.' In her view, such decisions were 'sending a powerful message to governments that a coal-fuelled economy is not compatible with a liveable planet' (Wood, 2015).

There has been much discussion as to whether Australia's Future Fund should follow the lead of Norway's sovereign wealth fund, and engage in fossil fuel divestment. In November 2014, Peter Costello – former treasurer, appointed chairman of the Future Fund – defended the decision of the Future Fund to allow for investments in respect of fossil fuels: 'I think it would be extraordinary if the government of Australia in its sovereign wealth fund said it was going to pull out of coal or gas or oil' (Yaxley, 2014). 'Go Fossil Free' has run a prominent campaign calling upon the Future Fund to invest in our future. The group contends that 'if we want to avoid dangerous climate change, the vast majority of fossil fuels must stay in the ground, yet the Future Fund is investing billions of dollars in coal, oil and gas companies' (Gofossilfree.org, 2015b). 'Go Fossil Free' has also highlighted investments of the Future Fund in controversial companies, such as BHP, Rio Tinto, Santos, Woodside, Chesapeake Energy and Gazprom, noting that it makes no sense for a future-focused institution to invest in companies like these. Instead, the Future Fund should follow the example of Norway's sovereign wealth fund (Gofossilfree.org, 2015b).

Conclusion

In the wake of the decision of the Government of Norway to engage in fossil fuel divestment, there has been a significant push for other national governments to follow suit with their sovereign wealth funds. A concerted effort to internationalise and globalise the fossil fuel divestment movement is a key feature. A similar case has been the explicit provisions about tobacco control divestment included in the World Health Organisation Framework Convention on Tobacco Control of 2003. But on climate change, there has been a hope that the 1992 *United Nations Framework Convention on Climate Change* (UNFCCC) could be revised and amended to include support for fossil fuel divestment by national governments.

With the Declaration on Climate Justice, the Mary Robinson Foundation (2013) called for fossil fuel divestment as a means of encouraging climate justice. This document highlighted the importance of investing in the future:

> A new investment model is required to deal with the risks posed by climate change – now and in the future, so that intergenerational equity can be achieved. Policy certainty sends signals to invest in the right things. By avoiding investment in high-carbon assets that become obsolete, and prioritizing sustainable alternatives, we create a new investment model that builds capacity and resilience while lowering emissions. Citizens are entitled to have a say in how their savings, such as pensions, are invested to achieve the climate future they want. It is critical that companies fulfil their social compact to invest in ways that benefit communities and the environment. Political leaders have to provide clear signals to business and investors that an equitable low-carbon economic future is the only sustainable option. (Mary Robinson Foundation, 2013)

Moreover, the Declaration on Climate Justice called for transformative climate leadership. The statement stressed: 'At the international level and through the United Nations (UN), it is crucial that leaders focus attention on climate change as an issue of justice, global development and human security' (Mary Robinson Foundation, 2013). Besides the work in the context of her Foundation, UN Climate Envoy Mary Robinson has also provided powerful support for the fossil fuel divestment movement (Rimmer, 2015b).

In 2014, Christiana Figueres – executive secretary of the UNFCCC – told the oil and gas industry: 'If we are to stay within 2 degree maximum temperature rise, and with the release of the new IPCC report this week, there is no doubt that we must, we have to, stay within a finite, cumulative amount of greenhouse gas emissions in the atmosphere.' She noted: 'We have already used more than half of that budget…this means that three quarters of the fossil fuel reserves need to stay in the ground, and the fossil fuels we do use must be utilized sparingly and responsibly.' (Figueres, 2014)

In September 2015, the UN lent its support to the Divest-Invest Campaign. Figueres emphasised: 'Investing at scale in clean, efficient power offers one of the clearest, no regret choices ever presented to human progress (UNFCCC, 2015). The UN secretary-general, Ban Ki-Moon, commented: 'I have been urging companies like pension funds or insurance companies to reduce their investments in coal and a fossil-fuel based economy to move to renewable sources of energy' (United Nations, 2014).

In March 2015, Nick Nuttall, the spokesman for the administration of the UNFCCC, commented: 'We support divestment as it sends a signal to companies, especially coal companies that the age of "burn what you like, when you like" cannot continue' (Carrington, 2015a). He provided support for the concept of the carbon budget: 'Everything we do is based on science and the science is pretty clear that we need a world with a lot less fossil fuels.' Nuttall stressed that the UN had lent its support to civil society and non-government organisations engaged in advocacy for fossil fuel divestment:

'We have lent our own moral authority as the UN to those groups or organisations who are divesting' (Carrington, 2015a). There is no doubt that there will be significant future debate as to whether international climate law should support fossil fuel divestment by nation states and sovereign wealth funds.'

We have lent our own moral authority as the UN to those groups or organisations who are divesting' (Carrington, 2015a). There is no doubt that there will be significant future debate as to whether international climate law should support fossil fuel divestment by nation states and sovereign wealth funds.

References

350.org (2013, 18 February). *Resolution Introduced to Push San Francisco to Divest from Fossil Fuels*. Press Release. Retrieved from http://www. motherearthnews.com/renewable-energy/fossil-fuel-divestment-zb0z1302zpit. aspx

350.org (2015, 27 October). *Melbourne City Council Commits to Fossil Free Investments Ahead of Paris Climate Talks*. Press Release. Retrieved from http://350.org.au/news/melbourne-city-council-commits-to-fossil-free-investing-ahead-of-paris-climate-talks/

ABC Environment (2015, 28 May). *Norway's Sovereign Wealth Fund Told to Divest from Coal Assets*. Retrieved from http://www.abc.net.au/environment/articles/2015/05/28/4244351.htm

Agence France-Presse (2015, 19 October). Oslo Moves to Ban Cars From City Centre Within Four Years. *The Guardian*. Retrieved from http://www.theguardian.com/environment/2015/oct/19/oslo-moves-to-ban-cars-from-city-centre-within-four-years

Ansar, A., Caldecott, B. and Tilbury, J. (2014). *Stranded Assets and the Fossil Fuel Divestment Campaign: What Does Divestment Mean for the Valuation of Fossil Fuel Assets?* Oxford: Oxford University Press. Retrieved from http://www.smithschool.ox.ac.uk/research-programmes/stranded-assets/SAP-divestment-report-final.pdf

Arabella Advisors (2015). *Measuring the Growth of the Global Fossil Fuel Divestment Movement and Clean Energy Investment Movement*. Retrieved from http://www.arabellaadvisors.com/wp-content/uploads/2015/09/Measuring-the-Growth-of-the-Divestment-Movement.pdf

Bolton, D. (2015, 7 October). Sweden Wants to Become the First Fossil Fuel-Free Country in the World – How Will It Work? *The Independent*. Retrieved from http://www.independent.co.uk/environment/sweden-first-fossil-fuel-free-country-in-the-world-a6684641.html

Carrington, D. (2015a, 15 March). Climate Change: UN Backs Fossil Fuel Divestment Campaign. *The Guardian*. Retrieved from http://www.theguardian.com/environment/2015/mar/15/climate-change-un-backs-divestment-campaign-paris-summit-fossil-fuels

Carrington, D. (2015b, 28 May). Norway Fund Could Trigger Wave of Large Fossil Fuel Divestments, Say Experts. *The Guardian*. Retrieved from http://www.theguardian.com/environment/2015/may/28/norway-fund-could-trigger-wave-of-large-fossil-fuel-divestments-say-experts?CMP=share_btn_tw

Carroll, R. (2015, 2 September). Coal Divestment Bill Passes California State Legislature. *Reuters*. Retrieved from http://www.reuters.com/

article/2015/09/02/us-california-divestiture-coal-idUSKCN0R226A20150902

City of Fremantle (2014, 13 November). *City of Fremantle to Divest from Carbon Intensive Investments*. News & Media. Retrieved from http://fremantle.wa.gov.au/news-and-media/city-fremantle-divest-carbon-intensive-investments

Cleverley, B. (2015, 26 July). Victoria Seeks Powers to Divest from Fossil Fuels. *Times Colonist*. Retrieved from http://www.timescolonist.com/news/local/victoria-seeks-powers-to-divest-from-fossil-fuels-1.2012468

Edis, T. (2015, 24 August). ACT Government to Divest from Fossil Fuels and Target 100% Renewable Energy. *The Australian Business Review*. Climate Spectator. Retrieved from http://www.businessspectator.com.au/news/2015/8/24/energy-markets/act-government-divest-fossil-fuels-and-target-100-renewable-energy

Figueres, C. (2014). *IPIECA 40th Anniversary Conference*, London, 3 April 2014. Retrieved from http://unfccc.int/files/press/statements/application/pdf/20140204_ipieca.pdf

Flannery, T. (2015). *Atmosphere of Hope: Searching for Solutions to the Climate Crisis*. Melbourne: The Text Publishing Company.

Free Speech Radio News (2015, 1 May). *New Zealand City to Divest from Fossil Fuels*. Retrieved from http://fsrn.org/2015/05/new-zealand-city-to-divest-from-fossil-fuels/

Gofossilfree.org (2015a). *Divest Norway*. Retrieved from http://gofossilfree.org/norway/Gofossilfree.org (2015b). *Divest the Future Fund*. Retrieved from http://gofossilfree.org.au/future-fund/

Goldenberg, S. (2015, 27 March). Rockefeller Brothers Fund: It Is Our Moral Duty To Divest From Fossil Fuels. *The Guardian*. Retrieved from http://www.theguardian.com/environment/2015/mar/27/rockefeller-fund-chairman-moral-duty-divest-fossil-fuels

Government of Norway (2014a, 4 April). The Norwegian Government Pension Fund Global – A financial Investor, not a Political Policy Tool. Retrieved from https://www.regjeringen.no/en/aktuelt/The-Norwegian-Government-Pension-Fund-Global---a-financial-investor-not-a-political-policy-tool/id755283/

Government of Norway (2014b, 9 April). *Company Exclusions*. Retrieved from https://www.regjeringen.no/en/topics/the-economy/the-government-pension-fund/internt-bruk/companies-excluded-from-the-investment-u/id447122/

Government of Norway (2014c, 18 December). *Guidelines for Observation and Exclusion from the Government Pension Fund Global*. Retrieved from https://www.regjeringen.no/globalassets/upload/fin/statens-pensjonsfond/guidelines-for-observation-and-exclusion-14-april-2015.pdf

Government of Norway (2015a, 10 April). *New Climate Criterion for the Exclusion of Companies from the Government Pension Fund Global (GPFG)*. Retrieved from https://www.regjeringen.no/en/aktuelt/nytt-klimakriterium-for-utelukkelse-av-selskaper/id2405205/

Government of Norway (2015b, 28 May). *The Government Pension Fund Global – Investments in Coal Companies*. Retrieved from https://www.regjeringen.no/en/aktuelt/the-government-pension-fund-global--investments-in-coal-companies/id2413829/

Green, M. (2013, 14 February). Bursting the Carbon Bubble *The Age*. Retrieved from http://www.theage.com.au/business/carbon-economy/bursting-the-carbon-bubble-20130214-2efob.html

Green, M. (2014, 15 September). Climate Activism's New Frontier is Targeting Fossil Fuel Investors. *The Age*. Retrieved from http://www.theage.com.au/national/climate-activisms-new-frontier-is-targeting-fossil-fuel-investors-20140912-10fxoc.html

Gunther, M. (2012, 30 November). Where Can Investors Who Worry About Climate Change Put Their Pension? *The Guardian*. Retrieved from http://www.theguardian.com/sustainable-business/blog/fossil-fuels-pension-divestment?INTCMP=SRCH

Ker, P. (2014, 4 March). Norwegian Wealth Fund May Ditch Australian Resource Firms. *The Sydney Morning Herald*. Retrieved from http://www.smh.com.au/business/carbon-economy/norwegian-wealth-fund-may-ditch-australian-resource-firms-20140303-340g0.html

Klein, N. (2014). *This Changes Everything: Capitalism vs The Climate*. New York: Simon & Schuster.

Klein, N. (2015, 17 October). Climate Change: How to Make the Big Polluters Really Pay. *The Guardian*. Retrieved from http://www.theguardian.com/commentisfree/2014/oct/17/climate-change-make-big-polluters-pay-fossil-fuel-industries

Law, S. (2015, 24 September). City, County Join World Fossil Fuels Divestment Movement. *Sustainable Life*. Retrieved from http://portlandtribune.com/sl/274399-150205-city-county-join-world-fossil-fuels-divestment-movement-

Leggett, J. (2014). *The Energy of Nations: Risk Blindness and the Road to Renaissance*. New York: Routledge.

Martin, C. (2015, 22 September). Fossil Fuel Divestment Movement Exceeds 2.6 Trillion. *Bloomberg Business*. Retrieved from http://www.bloomberg.com/news/articles/2015-09-22/fossil-fuel-divestment-movement-exceeds-2-6-trillion

Mary Robinson Foundation (2013, September 23). *Declaration on Climate Justice*. Retrieved from http://www.mrfcj.org/our-work/equity-and-climate-justice/declaration-climate-justice.html

Massie, R. K. (2012). *A Song in the Night: A Memoir of Resilience*. New York: Nan A. Talese/Doubleday.

McGinn, M. (2012, 22 December). Seattle Mayor Orders City to Divest from Fossil Fuels. *350.org*. Retrieved from http://350.org/seattle-mayor-orders-city-divest-fossil-fuels/

McKibben, B. (2012, 19 July). Global Warming's Terrifying New Math. *Rolling Stone*. Retrieved from http://www.rollingstone.com/politics/news/global-warmings-terrifying-new-math-20120719

McKibben, B. (2013). *Oil and Honey: The Education of an Unlikely Activist*. Melbourne: Black Inc. Books

McKie, R. and Helm, T. (2014, 12 April). UN Urges Huge Increase in Green Energy to Avert Climate Disaster. *The Guardian*. Retrieved from http://www.theguardian.com/environment/2014/apr/12/un-urges-increase-green-energy-avert-climate-disaster-uk?CMP=twt_gu

Mohsin, S. and Holter, M. (2015, 28 May). Norway's $1.15 Trillion Wealth Fund Curbs Coal Investments in New Industry Blow. *The Sydney Morning Herald*. Retrieved from http://www.smh.com.au/environment/climate-change/norways-115-trillion-wealth-fund-curbs-coal-investments-in-new-industry-blow-20150527-ghb8h1.html

Norges Bank Investment Management (2013). *Government Pension Fund Global: Annual Report 2013*. Retrieved from http://www.e-pages.dk/nbim/117/43

Norges Bank Investment Management (2015, 13 October). *Decision to Place Company in the Portfolio of the Government Pension Fund Global Under Observation*. Retrieved from http://www.nbim.no/en/transparency/news-list/2015/decision-to-place-company-in-the-portfolio-of-the-government-pension-fund-global-under-observation/

Phillips, A. (2014, 14 March). Norway's Sovereign Wealth Fund to Mandate Investment in Renewable Energy. *Climate Progress*. Retrieved from http://thinkprogress.org/climate/2014/03/14/3403251/norway-sovereign-wealth-fund-renewable-energy/

Picard, A. (2015, 26 August). Canadian Medical Association Divesting Fossil Fuel Holdings. *The Globe and Mail*. Retrieved from http://www.theglobeandmail.com/news/national/canadian-medical-association-divesting-fossil-fuel-holdings/article26115904/

Queally, J. (2015, 5 June). *Norway Goes Big on Fossil Fuel Divestment… Now Who's Next? Common Dreams*. Retrieved from http://www.commondreams.org/news/2015/06/05/norway-goes-big-fossil-fuel-divestment-now-whos-next

Reuters (2015, 28 May). Norway's $900bn Sovereign Wealth Fund Told to Reduce Coal Assets. *The Guardian*. Retrieved from http://www.theguardian.com/world/2015/may/27/norway-sovereign-fund-reduce-coal-assets

Richardson, B. J. (2013a). *Fiduciary Law and Responsible Investing: In Nature's Trust*. London: Routledge.

Richardson, B. J. (2013b). Sovereign Wealth Funds and Socially Responsible Investing: An Emerging Public Fiduciary. *Global Journal of Comparative Law*, 1(2), 125-162.

Rimmer, M. (2015a, 19 March). Can the Gates Foundation be Convinced to Dump Fossil Fuels? *The Conversation.* Retrieved from https://theconversation.com/can-the-gates-foundation-be-convinced-to-dump-fossil-fuels-38993

Rimmer, M. (2015b). Mary Robinson's Declaration of Climate Justice: Climate Change, Human Rights, and Fossil Fuel Divestment. In: H. Breakey, V. Popovski, and R. Maguire (eds). *Ethical Values and the Integrity of the Climate Change Regime* (pp. 189-212). Farnham, Surrey: Ashgate.

Rowling, M. (2015, 22 September). Actor DiCaprio Joins Growing Movement to Divest from Fossil Fuels. *Reuters.* Retrieved from http://www.reuters.com/article/2015/09/22/us-climatechange-energy-divestment-idUSKCN0RM2PZ20150922

Rubin, J. (2015). *The Carbon Bubble: What Happens to Us When It Bursts.* Toronto: Random House Canada.

Saunders, A. (2015, 26 August). Newcastle, Home of Coal, Joins Divestment Push. *The Sydney Morning Herald.* Retrieved from http://www.smh.com.au/business/mining-and-resources/newcastle-home-of-coal-joins-divestment-push-20150826-gj7xz1.html

Saxifrage, C. (2013, 11 February). University Divestment from Fossil Fuels are Taking Off – Municipalities are Next. *Vancouver Observer.* Retrieved from http://www.vancouverobserver.com/blogs/earthmatters/university-divestment-fossil-fuels-taking-municipalities-are-next

Schücking, H. (2015, 4 June). *Norway Divests!* Urgewald/Greenpeace/Framtiden. Retrieved from https://www.urgewald.org/sites/default/files/divestment.briefing_-_final.pdf

Shankleman, J. (2015, 14 October). Norwegian Prime Minister Demands Global Carbon Price and End to Fossil Fuel Subsidies. *Business Green.* Retrieved from http://www.businessgreen.com/bg/news/2430358/norwegian-prime-minister-demands-global-carbon-price-and-end-to-fossil-fuel-subsidies

UNFCCC (2015, 24 September). *Investors Worth $2.6 Trillion Looking to Divest Fossil Fuels: Divest-Invest Movement.* Press Release. Retrieved from http://newsroom.unfccc.int/financial-flows/fossil-fuel-divestment-pledges-surpass-26-trillion/

United Nations (2014, 4 November). *Ban Ki-Moon Urges More Fossil Fuel Divestment*. UN Newsroom. Retrieved from http://newsroom.unfccc.int/ financial-flows/ban-ki-moon-speaks-in-favour-of-divestment/

Vaughan, A. (2014, 9 October). Lego Ends Shell Partnership Following Greenpeace Campaign. *The Guardian*. Retrieved from http://www. theguardian.com/environment/2014/oct/09/lego-ends-shell-partnership-following-greenpeace-campaign

Wood, C. (2015, 10 July). Coal Divestment is Just the Beginning – Let's Not Forget Oil and Gas. *Huffington Post*. Retrieved from http://www. huffingtonpost.com/charlie-wood/coal-divestment-is-just-t_b_7766310.html

WWF Norge (2014, 4 April). *Norway Raises Ambition, But Not the Bar on Renewable Energy*. Retrieved from http://wwf.panda.org/wwf_news/?219116/ Norway-raises-ambition-but-not-the-bar-on-renewable-energy

Yaxley, L. (2014, 19 November). Peter Costello, Chairman of the Future Fund, Defends Investment in Fossil Fuels after Divestment of Tobacco Shares. *The World Today*. ABC News. Retrieved from http://www.abc.net.au/ news/2014-11-20/peter-costello-defends-the-future-fund27s-fossil-fuels-investm/5906458

Conclusions
Looking Beyond International Relations

GUSTAVO SOSA-NUNEZ

INSTITUTO MORA, MEXICO

&

ED ATKINS

UNIVERSITY OF BRISTOL, UK

There was 'No Plan B' read the illuminated display on the Eiffel Tower during the 2015 United Nations Framework Convention on Climate Change (UNFCCC) Conference of Parties (COP-21) in Paris, France. Due to the collapse of four previous annual UNFCCC conferences (Copenhagen 2009, Cancun 2010, Durban 2011 and Doha 2012), the spectre of climate change still remained an unsolved problem at the top of the international agenda. Despite the political atmosphere becoming dominated by the Islamic State's attacks on Paris on 13 November and the associated shutdown of protests outside Le Bourget, the promise of the Paris summit was tangible – with delegates working through days and nights in an attempt to reach an agreement that was accepted by all.

The numbers are there to show the significance of Paris as a turning point in the climate change regime. Over 150 world leaders descended upon Paris for the opening day of the summit and 180 nations put forward plans how, and by how much, they will curb carbon emissions by 2030. The usual roadblocks to progress were there – questions of temperature goals (1.5 degrees Celsius or 2 degrees Celsius), decarbonisation goals of zero emissions, financial mechanisms of support, the expected loss and damage to be caused by climatic change, debates of development, and the need for future improvements all dominated the 2015 negotiations. Yet, these debates did not result in controversy or political failure. There were limited reports of the abandonment of constructive dialogue and discussion and the reports from the conference continued to be positive, as the event continued into its final days and hours. It would seem that the delegates heeded the words of Christina Figureres, executive secretary of the UNFCCC, that 'The Paris Agreement is not only possible, it is necessary and urgent. We are counting on everyone's contribution' (UNFCCC, 2015).

As the delegates gathered at Le Bourget on Saturday 12 December, the atmosphere was one of positivity. The arrival of French president Francois Hollande was greeted with applause. Members of the High Ambition Coalition provided an important symbolism by entering the room together. Negotiators from the bloc of Least Developed States voiced optimism about the draft text published hours before. As United Nations (UN) secretary general Ban Ki Moon stated at the opening of the plenary,

> We have come to a defining moment on a long journey that dates back decades. The document, with which you have just been presented, is historic. It promises to set the world on a new path, to a low emissions climate resilient future. Let us now finish the job. The whole world is watching. (Vaughan and Randerson, 2015)

It is important to note that this collection was written in the weeks and months preceding the 2015 Paris Summit. For this reason it has not attempted to directly engage with the debates surrounding the conference at Le Bourget. Instead, this collection aims to provide an important introduction to the intricate linkages between climate change and international governance and its numerous facets, theories and nuances. Although COP-21 has created an important juncture in the regime of climate change mitigation and adaptation, it does not detract from the relevance of many of the contributions to this collection – that have aimed to provide different insights about the role that the environment and climate change play in terms of International Relations (IR).

In doing so, each contribution has provided an important understanding of how the Paris agreement – although it may provide an ambitious, balanced and historic moment – cannot be the climax. It is the start of global efforts, not the end. World societies need to work together for assembling a comprehensive approach to environmental issues that can only improve future environmental and climate conditions. Irrespective of whether we are – or become – *defensive* or *altruistic* environmentalists (Rudel, 2013: 5); we need to take additional actions in the immediate term. It is this statement that the collection has taken as its starting point – by providing a number of complimentary reviews and understanding of the place of climate change within the theories and practices of international governance. This conclusion shall now explore the perspectives provided – drawing links between the chapters and the viewpoints evident within them.

The first section of this edited collection explored the tendencies, the background and the context in which the international community is found.

Recognising the relevance of the environment to contemporary society and how we are inducing climate change is important. The linkages between the two are co-constitutional – with patterns of international politics and economy often creating the processes behind climate change, which in turn influence the nature of global governance. As a result, we must seek to understand how climate change is not only present in the nature of International Relations – but how it constitutes it.

In his contribution, Mizan R. Khan reviewed the main strands of International Relations theory – such as realism, liberalism and constructivism – as a window to understand the contemporary approaches to climate change adaptation. In doing so, Khan forwards an important framework to understand the current climate change regime as an amalgamation of theoretical strands of neoliberalism, regime theory and institutional functionalism. However, this is not assured. Instead, it is important to understand the creation of a new norm – that of climate change as a *global public good/global public bad* – a norm particularly evident in the Paris COP-21 negotiations. Khan's use of the *global public good/global public bad* framework allows for the opening up of analysis to the links between climate change and the fate of the world's most vulnerable peoples – as pursued in Úrsula Oswald Spring's contribution. With the ties between climate change and societal vulnerabilities evident, this contribution asserts that global policy must seek to promote sustainable actions and individual agency, while improving the livelihood of vulnerable human beings should be of utmost importance. For the author, such a policy must include the decarbonisation of the economy, energy efficiency, renewable energy use, reforestation and restoration of ecosystems. This, of course, should entail adjustments to the existing model of civilisation.

While Oswald Spring discussed the importance of agency, Simon Dalby explored the role that the environment is acquiring in International Relations scholarship. Dalby considers that this role is increasing – although it has not yet been sufficiently related to main IR issues such as war, peace and security. The logical option for the solidifying of these linkages would be through international regimes that characterise environmental matters. Yet, delays to implement environmental regulations, industrial evasions and campaigns of obfuscation and denial have obstructed efforts. Hence, a radical rethinking of the role of the state within the framework of international governance is necessary.

However, in the shadow of Paris 2015, it is possible to assert of the importance that the Intergovernmental Panel on Climate Change (IPCC) is having in world affairs – due to its promotion of climate change as anthropogenic. It is this assertion that Nina Hall has pursued in her contribution. For this, Hall has provided a detailed account of the increasing importance that climate change has acquired in recent years, to the point of being at the top of the agenda at certain international gatherings, such as the G7 and the G8. The reason for this can be found in how climate change is now more than an environmental issue. It has become an economic and security concern.

An important barrier to the success of climate change mitigation and adaption can often be found in the unaccommodating perceptions of individuals – be it in the form of climate-scepticism or risk aversion. Whether it is by ignorance, scientific misunderstandings, political convictions or intended misinformation, the contribution of Kirsti Jylhä asserted that denial plays an important part of how public perception on climate change is at present. Jylhä posits that such denial is often caused and fuelled by uncertainty, fear and doubt. By identifying that the majority of the literature providing counter-evidence for climate change is published outside scientific communities and has links to politically conservative movements, she suggests that political orientation is a core aspect to consider in moving forward.

The second section aimed to explore the routes forward in building a comprehensive understanding of the interactions between climate change and international governance. A number of divergent views on policies, security, finance and non-governmental actors' roles were presented, illustrating that change on current trends is possible. However for this transformation to occur, the participation of diverse actors and a change in perspective is necessary.

The important nature of environmental policies as transversal was addressed by Gustavo Sosa-Nunez. In his view, environmental perspectives should be of mandatory presence across broader policy frameworks, especially in cases when transboundary environmental problems occur. Yet, although environmental policies are being increasingly considered within policy frameworks, this often occurs slowly and, in many cases, on the periphery of wider policy regimes. The reason for this can be found in complexity. Grand projects, policies and actions – to be developed across wider geographical settings – tend to be more difficult to achieve. The more parties are involved, the more complicated it is to reach consensus or agreement on the path to follow. It is this problem of complexity and the overlapping of environmental issues with traditional political concerns that forms the foundations of Ed Atkins' analysis of environmental conflict. This contribution looked to debunk traditional neo-Malthusian assertions of the links between climate change and violent conflict by exploring a number of significant limitations in such an argument. In many cases of conflict (such as World War II, the Iraqi invasion of Kuwait and the 1969 Soccer War), environmental factors often interact with social, political and economic issues to create situations of conflict. However, this contribution argued that it is often impossible to separate this web of causation. As a result, asserting that a conflict is environmental is deterministic and neglects additional causes. This creates a dangerous situation for the analysis of contemporary and future conflict.

The assertions of the contributions by Sosa-Nunez and Atkins both point to the need for an important shift in the framing of socio-natural interactions – one that may provide non-governmental organisations (NGOs) with a new window of impact on policy. In aiming to identify the extent to which civil society and NGOs are drivers of change, Emilie Dupuits has shown the importance of considering the plurality of non-governmental actors in environmental terms. The increased involvement often widens the spectre of participating actors. Yet, this situation may also lead to the increased competition for power between the stakeholders involved – resulting in important trade-offs between demands. A potential outcome of this could thus be the hampering of the role that non-state actors play in environmental policy processes. As a result, this contribution proposes to reframe strategies so as to further civil society's participation within global environmental governance.

This need for transformation can be found in the final contribution to this section. In this contribution, Simone Lucatello has shown that environmental aid effectiveness depends upon a number of features. With this, variables often resulting in a multilateral approach to aid work far better than bilateral schemes. However, current trends continue to demonstrate a preference for bilateral assistance and for support for larger developing countries rather than those smaller nations with limited response capacities. Lucatello illustrates this argument with the case of Latin America – but this situation can be replicated elsewhere. The post-COP-21 regime may assist to this goal. However, if the current proliferating pattern that finance mechanisms persists, challenges to adequately allocate aid with environmental purposes will also continue.

The collection's final section has shown that the governance of climate change must move beyond the ivory tower of academia and the hallowed halls of politics and diplomacy – moving towards the commitments of institutions and individuals. In doing so, these contributions provide a brief introduction to how these varying settings can interact to provide an important route forward – while pointing to a complex but promising future.

As Lau Blaxekjær explored in his contribution, environmental diplomacy is enjoying resurgence within the sphere of the international climate change regime. Using the cases of the role of the Cartagena Dialogue in UNFCCC negotiations and the 3GF and other green growth networks, this contribution charted how diplomatic networks have allowed the translation of local, national and regional concerns into the language of international governance. As described in his postscript on the 2015 Paris negotiations, these diplomatic routes have started to bear some important fruit due to their anchoring of the issue-linkage and coalition building within the COP-21 negotiations. Significantly, this represents an important transition to the role of *partnerships* in both the practice and theory of environmental diplomacy. It is these partnerships that allow for the entrance of new communities into the international relations of climate change.

Yet, as Duncan Depledge has argued in his study of the geopolitics of the Arctic region, decisions surrounding climate change occur at all levels of governance – from the global to the local community. Within this context, the priorities of communities are often pitted against the interests of other international groups. Two possible futures are posited – with the 'opening up' of the Arctic region to international processes conflicting with the continued 'saving' of the area. These futures are formed of coalitions of local, domestic and international actors committed to divergent storylines of what the Arctic region can provide to the globe and its role within wider processes of climate change governance. As Depledge argues, how this competition plays out will have important consequences across the political scale. It is this complexity of local–global interaction that provides the starting point of Lada V. Kochtcheeva's analysis of the implementation of renewable energy strategies – a contribution that possesses links with many of the views previously voiced within the collection. For Kochtcheeva, the success of renewable energy schemes is often constrained by barriers created by other policy goals – be they of an economic, technological, regulatory or social nature. Evidence for this can be found in the continuation of subsidies for fossil fuel or nuclear industries – often contrasted with the limited support for more renewable energy regimes. Although the energy may be renewable and eternal, it is often evident that the political and financial support for them are not – as illustrated by the Conservative UK government's decision to end such support in 2015. As a result, Kochtcheeva argues for the adoption of a more systematic approach in academic literature – to understand the complex, and inherently political, barriers to the adoption of these energy technologies at the national level.

However, it is important to note that these institutional barriers can be – and are being – broken down. The final two contributions of this collection explore the popular divestment movement that has the mantra that you cannot solve the problem by supporting the actors that created it at its heart. In their contribution, Leehi Yona and Alex Lenferna look to the roots of this movement – educational institutions – to understand the future routes forward. It is important to note that the divestment movement is relatively youthful – starting in the spring of 2010 at Swarthmore College, Pennsylvania. However, by September 2014, over 500 institutions had committed to divest – totalling over US\$3.4 trillion in assets. Yona and Lenferna argue that the reasons behind this explosion of divestment activity can be found in the seizing of the torch by a new generation that has concentrated on tactics of issue linkage, support and pressure from a number of different interest groups and the belief that change is impossible. As Matthew Rimmer argues, an important consequence of this popular movement has been the increased focus on the need for sovereign wealth funds of nations to divest from the fossil fuel industries. Using a number of primary sources, Rimmer has explored this pressure and process within the case of Norway's 2015 decision to divest its US\$900 billion sovereign wealth fund from coal industries (Carrington, 2015). As this contribution argues, this decision has important consequences for the international system – as the result of individual and institutional commitments to the routes forward in climate change governance. Yet, this introduction of divestment as a policy tool is still new and faces a number of barriers to its successful implementation. However, these complementary contributions point to a more promising future; one in which the uphill struggle against fossil fuels can be countered by activity on the ground.

As we write this, the ink on the Paris Agreement of 2015 is barely dry. The promise is there. Yet, that is the issue. These are promises, not actions. At this stage, it is not known how successful the post-Paris regime will be. As this book has outlined, more must (and can) be done. The success of the climate change regime cannot be found in an agreement exclusively. With no Plan B, Plan A must involve a degree of reappraisal, action at a number of levels and the understanding that climate change is more than just Paris.

References

Carrington, D. (2015, June 5). Norway Confirms \$900bn Sovereign Wealth Fund's Major Coal Divestment. *The Guardian*. Retrieved from http://www.theguardian.com/environment/2015/jun/05/norways-pension-fund-to-divest-8bn-from-coal-a-new-analysis-shows

Rudel, T. K. (2013). *Defensive Environmentalists and the Dynamics of Global Reform*, Cambridge: Cambridge University Press.

UNFCCC (2015). P*aris Climate Change Conferenc*e – November 2015. Retrieved from http://unfccc.int/meetings/paris_nov_2015/meeting/8926.php

Vaughan, A. and Randerson, J. (2015, December 12). Paris Climate Talks: Governments Adopt Historic Deal – As It Happened. *The Guardian*. Retrieved from http://www.theguardian.com/environment/live/2015/dec/12/paris-climate-talks-francois-hollande-to-join-summit-as-final-draft-published-live

Contributors

Lau Øfjord Blaxekjær holds a PhD in Political Science from the University of Copenhagen. Currently, he researches and teaches at the University of the Faroe Islands, where he is Assistant Professor and Programme Director of the Masters in West Nordic Studies, Governance and Sustainable Management – a joint Master's Degree with participating universities from Greenland, Iceland and Norway. He has also been visiting professor at the Sino-Danish Centre for Education and Research in Beijing, China. His research interests include global governance, climate change, energy, risk society, Anthropocene, green growth, and theories of power, rhetoric and narrative as well as research and education in a transdisciplinary perspective. He is co-author of 'Mapping the narrative positions of new political groups under the UNFCCC', in *Climate Policy* (2014, with T. D. Nielsen).

Simon Dalby is CIGI Chair in the Political Economy of Climate Change at the Balsillie School of International Affairs and Professor of Geography and Environmental Studies at Wilfrid Laurier University, Waterloo, Ontario. He was educated at Trinity College Dublin, the University of Victoria, and holds a PhD from Simon Fraser University. Dalby is the author of *Environmental Security* (University of Minnesota Press, 2002) and *Security and Environmental Change* (Polity, 2009) and recently co-edited (with Shannon O'Lear, at the University of Kansas) *Reframing Climate Change: Constructing Ecological Geopolitics* (Routledge, 2016).

Duncan Depledge completed his PhD at Royal Holloway, University of London in 2014, with a focus on the development of contemporary British policy towards the Arctic from a critical geopolitical perspective. He currently combines a post-doctoral teaching fellowship at Royal Holloway with his role managing the secretariat to the British All-Party Parliamentary Group for Polar Regions in Westminster. Previously, he also held a research position (2009-2014) on the Climate Change and Security Programme at the Royal United Services Institute.

Emilie Dupuits is a PhD Candidate and Teaching Assistant at the Global Studies Institute and the Department of Political Science and International Relations at the University of Geneva, Switzerland. Her research interests focus on the governance of common natural resources – such as water and forests – in Latin America and on transnational networks emerging from the civil society. Her thesis is titled *From Grassroots Organizations to Transnational Networks: The Transformations of Water and Forests Community Governance in Latin America.*

Nina Hall is a Post-Doctoral Fellow at the Hertie School of Governance, in Berlin, Germany. Her research explores how international organisations are evolving in the 21st century. She has published on climate change, humanitarianism and gender equality in: *Global Environmental Politics, Global Governance,* and the *Australian Journal of Political Science.* She has a forthcoming book with Routledge, *Displacement, Development and Climate Change: International Organizations Moving beyond their Mandates.* Hall completed a PhD in International Relations at the University of Oxford. She has previously worked for the New Zealand Ministry of Foreign Affairs and for short periods with UNICEF Nepal and the UN Department of Political Affairs.

Kirsti M. Jylhä (née Häkkinen) is a PhD Candidate in Psychology at Uppsala University, Sweden. Her research investigates how and why certain socio-political ideologies (social dominance orientation, liberal-conservative/left-right political orientation, system justification and right-wing authoritarianism) are related to environmental attitudes and behaviour. Her thesis focuses on the relation between social dominance orientation and climate change denial.

Mizan R. Khan is currently a Professor of Environmental Management at North South University, Dhaka, Bangladesh. He received his PhD from the School of Public Policy, University of Maryland, USA, and has previously held positions at Brown University, University of Manitoba, Université de Poitiers and the University of Calcutta. His recent work consists of a wide range of publications on environmental issues and climate change, including two recent books: *Towards a Binding Climate Change Adaptation Regime: A Proposed Framework* (London & New York: Routledge, 2014, 2015) and *Power in a Warming World: The New Politics of Climate Change and the Remaking of Environmental Inequality* (Cambridge: MIT Press, 2015) – the latter co-authored with David Ciplet and J. Timmons Roberts. In addition to his academic work, he has formed part of the Bangladeshi delegation to IPCCC negotiations since 2001.

Lada V. Kochtcheeva received her MS and PhD in Political Science from the University of Oregon. She is currently an Assistant Professor of Political Science in the School of Public and International Affairs at North Carolina State University. Her current research is inspired by contemporary problems of environmental governance, institutional development, government–society relations and comparative public policy. She has published *Comparative Environmental Regulation in the United States and Russia: Institutions,*

Flexible Instruments, and Governance (New York: SUNY Press, 2008) – which examines political and policy pre-conditions for the replacement of command and control systems with flexible instruments such as incentive programmes, tradeable permits and pollution charges.

Alex Lenferna is a South African Mandela Rhodes and Fulbright Scholar at the University of Washington, conducting his PhD focusing on climate justice. His research has examined a range of related topics, including climate change-induced migration, geoengineering, carbon taxes, climate reparations, global poverty and the rights of nature. He is heavily involved in Divest University of Washington, the Gates Foundation Fossil Fuel Divestment campaign and the 350.org Seattle City Employees Retirement System Divestment Campaign. Drawing on his research and advocacy work, he is currently writing a book on the fossil fuel divestment movement and climate justice. He is also on the steering committee of Carbon Washington, a non-profit advocating for a progressive revenue-neutral carbon tax in Washington State.

Simone Lucatello is a Full-time Researcher at Instituto Mora, a public research centre based in Mexico City. He is the Director of the Research Programme in International Cooperation, Development and Public Policy at the same institution. Lucatello holds a joint BA in History from the University of Venice Cá Foscari, Italy, and University College London (UCL). He also holds an MA in International Relations from the London School of Economics and Political Science (LSE) and a PhD in Governance for Sustainable Development from Venice International University (VIU), Italy. His research interests deal with disaster relief, climate change, humanitarian action and sustainability. Lucatello has worked as consultant to many United Nations agencies, including UNEP, UNODC, UNIC and OCHA, as well as the Inter-American Development Bank (IADB). He is author of more than ten books and many articles published in English, Spanish, and Italian. He currently coordinates the Mexican Research Network on International Cooperation and Development (REMECID).

Úrsula Oswald Spring is Professor at the Regional Centre for Multidisciplinary Research of the National Autonomous University of Mexico, focusing on gender and equality. She holds a PhD in Social Anthropology with a focus on Ecology from the University of Zurich. She was the first Chair for Social Vulnerability of the United Nations University (UNU-EHS). She is a member of the Intergovernmental Panel on Climate Change (IPCC), Working Group II, and of the World Social Science Report. Oswald Spring has been reviewers' coordinator on water issues in the context of the Global Environmental Outlook (GEO-5). She was Minister for Environmental Development (1994–1998) in the Mexican State of Morelos, and the first female Ecology Attorney in Mexico (1992–1994). Moreover, she was also President of the International Peace Research Association (1998–2000) and Secretary-General of the Latin American Council for Peace Research (2002–2006).

Matthew Rimmer (BA/LLB ANU, PhD UNSW) is a Professor in Intellectual Property and Innovation Law at the Faculty of Law in the Queensland University of Technology (QUT). Rimmer is a leader of the QUT Intellectual Property and Innovation Law Research Program and a member of the QUT Digital Media Research Centre (QUT DMRC), the QUT Australian Centre for Health Law Research (QUT ACHLR) and the QUT International Law and Global Governance Research Program. Rimmer would like to acknowledge Charlotte Wood for her conversations, research and public policy work on future funds and fossil fuel divestment. This research was supported by an Australian Research Council Future Fellowship on *Intellectual Property and Climate Change: Inventing Clean Technologies*.

Leehi Yona is pursuing a BA in Environmental Studies and Biology (Minor Public Policy) at Dartmouth College in Hanover, New Hampshire. Yona has studied Arctic science and policy as a Penelope W. and E. Roe IV Stamps Scholar and James O. Freedman Presidential Scholar at Dartmouth. Yona is also a community organiser and has worked on several youth climate justice campaigns, including Power Shift Canada and USA and the Divest Dartmouth fossil fuel divestment campaign. Yona is a recipient of the Lieutenant Governor of Québec's Youth Medal and was named Canada's Top Environmentalist under 25 in 2013.

Note on Indexing

E-IR's publications do not feature indexes due to the prohibitive costs of assembling them. However, if you are reading this book in paperback and want to find a particular word or phrase you can do so by downloading a free e-book version of this publication in PDF from the E-IR website.

When downloaded, open the PDF on your computer in any standard PDF reader such as Adobe Acrobat Reader (pc) or Preview (mac) and enter your search terms in the search box. You can then navigate through the search results and find what you are looking for. In practice, this method can prove much more targeted and effective than consulting an index.

If you are using apps such as iBooks or Kindle to read our e-books, you should also find word search functionality in those.

You can find all of our e-books at: http://www.e-ir.info/publications

Printed in Great Britain
by Amazon

66918117R00142